Augmented Vision and Reality

Volume 6

Series Editors

Riad I. Hammoud, Kokomo, IN, USA
Lawrence B. Wolff, New York, NY, USA

For further volumes:
http://www.springer.com/series/8612

Vijayan K. Asari
Editor

Wide Area Surveillance

Real-time Motion Detection Systems

Springer

Editor
Vijayan K. Asari
Electrical and Computer Engineering
University of Dayton
Dayton, OH
USA

ISSN 2190-5916 ISSN 2190-5924 (electronic)
ISBN 978-3-642-37840-9 ISBN 978-3-642-37841-6 (eBook)
DOI 10.1007/978-3-642-37841-6
Springer Heidelberg New York Dordrecht London

Library of Congress Control Number: 2013950723

Printed on acid-free paper

Springer is part of Springer Science+Business Media (www.springer.com)

Preface

Wide area surveillance refers to an automated monitoring process that involves data acquisition, analysis, and interpretation for understanding object behaviors. Automated surveillance systems are mostly used for military, law enforcement, and commercial applications. Sensors of different types and characteristics in surface-based or aerial-based platforms are used for the acquisition of data of large areas sometimes covering several square miles. Intelligent visual surveillance is becoming more popular in applications such as human identification, activity recognition, behavior analysis, anomaly detection, alarming, etc. Detection, tracking, and identification of moving objects in a wide area surveillance environment have been an active research area in the past few decades. Object motion analysis and interpretation are integral components for activity monitoring and situational awareness. Real-time performance of these data analysis tasks in a very wide field of view is an important need for monitoring in security and law enforcement applications.

This edited book, *Wide Area Surveillance: Real-time Motion Detection Systems*, aims to present a few selected state-of-the-art research and development outcomes pertaining to real-world applications. It is part of a Springer Book series titled *Augmented Vision and Reality* which has been initiated by Dr. Riad Hammoud. This book includes a wide variety of active research topics which are very relevant to wide area surveillance for scene analysis and understanding. The major research areas addressed in the 10 chapters of this book are background elimination, shadow detection, moving object detection and tracking, human activity recognition and crowd monitoring, person identification, networked camera controls, and target recognition in multi-spectral and hyper-spectral imagery.

The first part of the book covers research on wide area surveillance data analysis for background subtraction, moving cast shadow detection, and moving object detection and tracking. The chapter on *Background Subtraction* by Ahmed Elgammal of Rutgers University presents detailed description of various statistical approaches to model scene background. An extensive review of the concept and the practice in background subtraction is provided in this chapter to establish the need for this important research activity in wide area surveillance and scene understanding. Several basic statistical background subtraction models such as parametric Gaussian models and nonparametric models are presented. It also discusses the issue of shadow suppression for human motion analysis applications and different approaches and tradeoffs for background maintenance. Many recent developments in background subtraction paradigm are also addressed in this chapter.

The chapter on *Moving Cast Shadows Detection* by Ariel Amato, Ivan Huerta, Mikhail G. Mozerov, F. Xavier Roca, and Jordi González of Universitat Autònoma de Barcelona reviews several shadow detection methods as well as their taxonomies related to change detection, moving object detection, scene matching, and visual surveillance in a background subtraction context. Detection of moving cast shadows is an important research activity in a broad range of vision-based surveillance applications as it helps in improving the performance of long-range object classification tasks. Several shadow detection methods are presented in this chapter to work with static images and with image sequences. The detected shadow regions in an image can be exploited to obtain geometric and semantic cues about shape and position of its casting object and localization of the light source. The shadows in video sequences can be exploited for detecting changes in a scene and object matching. The effect of shadows on the shape and color of the target object which may affect the performance of scene analysis and interpretation is also discussed in this chapter. The chapter on *Object Detection and Tracking in Wide Area Motion Imagery* by Varun Santhaseelan and Vijayan K. Asari of University of Dayton presents new methodologies for moving object detection and feature-based object tracking in low resolution video that can work well in varying lighting conditions as well as for very small objects like pedestrians in an aerial image. The basic philosophy employed in the development of the new algorithm is that the entire information available in an object of interest has to be utilized in the detection and tracking processes. A dense version of a localized histogram of gradients on the difference images forms the feature set for representing a low resolution object region. The effectiveness of a nonlinear image enhancement and super-resolution algorithms in relationship with tracking in shadows is also presented in this chapter.

The second part of the book covers research and development of algorithms for human activity recognition, autonomous crowd monitoring, and human identification in a wide area surveillance environment. The chapter on the *Recognition of Complex Human Activities in a Crowd Context* by Wongun Choi and Silvio Savarese of University of Michigan examines the problem of classification of collective human activities from video sequences. The presence of the coherent behavior among individuals in a spatial and temporal neighborhood is defined as a collective activity in this chapter. Collective activities such as queuing in a line or talking are defined by observing the interactions of nearby individuals in time and space. Several recent methods for analyzing collective activities through the concept of crowd context are presented in this chapter. Various solutions for modeling the crowd context are also discussed along with demonstration of the flexibility and scalability of the new framework on several datasets of collective human activities. The chapter on *Autonomous Cognitive Crowd Monitoring* by Simone Chiappino, Lucio Marcenaro, Pietro Morerio, and Carlo Regazzoni of University of Genoa presents an event-based dynamic Bayesian network that can switch among alternative Bayesian filtering and control lower level modules to capture adaptive reactions of human operators. An intelligent video surveillance system should be able to represent complex situations that describe actions and activities of humans in a dynamically varying background. Human behavior analysis is an important research activity in cognitive crowd monitoring. A cognitive decision making process includes an automatic support to human decisions based on object detection, tracking, and situation assessment in the environment. The new event-based switched dynamic Bayesian network presented in this chapter can be used to represent and anticipate possible actions and activities within the intelligent video surveillance context. This can also interact with an active visual simulator of crowd situations.

The chapter on *Unified Representation of Human Faces for Recognition of Individuals in Surveillance Videos* by Le An, Bir Bhanu, and Songfan Yang of University of California, Riverside uses a novel image representation of a unified face image which is synthesized from multiple camera video feeds. Low quality of the probe data in terms of resolution, noise, blurriness, and varying lighting conditions and the variations in face poses make the recognition of individuals in a video surveillance environment more difficult and less accurate in real-world surveillance video data captured in a multi-camera network. The unified face image representation warps the probe frames from different cameras toward a template frontal face which then generate a frontal view of the subject that incorporates information from different cameras. The unified face image representation framework presented in this chapter is a generalized approach which can be adapted to any multi-camera video-based face recognition system that uses any face feature descriptors and classifiers. The chapter on *Person Re-identification in Wide Area Camera Networks* by Shishir K. Shah and Apurva Bedagkar-Gala of University of Houston considers the context of consistent people tracking over multiple cameras in order to facilitate the estimation of the global trajectory of a person over the camera network. Person re-identification is a fundamental task in wide area surveillance for multi-camera tracking and the subsequent analysis of long-term activities and behaviors of people in the scene. The appearance of an individual in a surveillance environment may be affected by varying background lighting, poses, and orientations, and camera to person distance. These cause the deterioration of accuracies in feature extraction and classification processes. This chapter discusses these issues and presents a re-identification model that works well in such challenging conditions. A multi-parametric model and its effectiveness for person re-identification are also presented in this chapter.

The third part of the book covers research on navigation aid for unmanned air systems, automatic target recognition on multi-spectral and hyper-spectral imagery, and distributed sensor data processing. The chapter on *Opportunities and*

Challenges of Terrain Aided Navigation Systems for Aerial Surveillance by Unmanned Aerial Vehicles by Samil Temel and Numan Unaldi of Turkish Air Force Academy presents the development of a terrain aided navigation system and its use as a test-bed for the design of an autonomous navigation system. Unmanned aerial vehicles are now becoming popular in military and civilian surveillance applications as they provide more accurate, inexpensive, and durable information than ground surveillance systems. Unmanned aerial surveillance systems gather data using various sensors equipped on them. Most of the current unmanned aerial systems depend on satellite-based navigation systems which are likely to be jammed in military fields. The terrain aided navigation system presented in this chapter provides position estimates relative to known terrains by utilizing the height values from the surface with the help of active range sensors which are then matched within a terrain digital elevation map. This chapter also summarizes some of the design objectives for unmanned aerial vehicles-based surveillance posts.

The chapter on *Automatic Target Recognition in Multi-Spectral and Hyper-Spectral Imagery* by Mohammad S. Alam and Adel Sakla of University of South Alabama discusses the one-dimensional spectral fringe-adjusted joint transform correlation-based technique for detecting very small targets involving only a few pixels in multi-spectral and hyper-spectral imagery. The joint transform correlation of the spectral signatures from the unknown hyper-spectral imagery with the reference signature can detect both single and multiple desired targets in constant time while accommodating the in-plane and out-of-plane distortions. The proposed joint transform correlation technique is also applied to the discrete wavelet transform coefficients of the multi-spectral and hyper-spectral data in order to improve the detection performance. This chapter also presents the effectiveness in performance of the proposed method in some real-life hyper-spectral image data cubes. The chapter on *Distributed Estimation and Control in Camera Networks* by A. Kamal, C. Ding, A. Morye, J. A. Farrell, A. Roy-Chowdhury of University of California, Riverside reviews some of the state-of-the-art techniques in distributed computer vision algorithms related to distributed estimation and distributed control scenarios. Distributed processing is a necessity in several application domains such as national security, home monitoring, and environmental monitoring where large-scale camera networks are deployed. The limitations of setting up a substantial communication infrastructure beforehand and the possible mobility of the sensors may also necessitate a distributed processing environment. The basic consensus algorithms and analysis of their applicability in camera networks are discussed in the distributed estimation section and presents some modifications that would consider the constraints posed by vision sensors. A review of game-theoretic cooperative control algorithms is presented in the distributed control section and proposes how they can be adapted for active sensing in a camera network that leads to an integrated sensing and control paradigm.

The Editor and the Authors of this book believe that it provides a good source of information and references for academic and industrial researchers, professionals in intelligence, surveillance and reconnaissance (ISR) community, and graduate and senior undergraduate students for conducting wide area image analysis and scene understanding research. The Editor thanks all the Authors for their contributions, Dr. Riad Hammoud for proposing this project, and the reviewers for their invaluable service in making this project a great success.

October 2012 Vijayan K. Asari

Contents

Editor Biography

Dr. Vijayan K. Asari is a Professor in Electrical and Computer Engineering and Ohio Research Scholars Endowed Chair in Wide Area Surveillance at the University of Dayton, Dayton, Ohio. He is the Director of the University of Dayton Vision Lab (Computer Vision and Wide Area Surveillance Laboratory). Dr. Asari had been a Professor in Electrical and Computer Engineering at Old Dominion University, Norfolk, Virginia till January 2010. Dr. Asari was the Founding Director of the Computational Intelligence and Machine Vision Laboratory (ODU Vision Lab) at ODU. His current research areas include image and video processing for object detection, tracking and identification, wide area surveillance and scene analysis for security automation and situational awareness, face detection and recognition, facial expression analysis, human action and activity recognition, brain signal analysis for emotion recognition and brain machine interface. Dr. Asari received the Bachelor's degree in Electronics and Communication Engineering from the University of Kerala (College of Engineering, Trivandrum), India, in 1978, the M.Tech and Ph.D. degrees in Electrical Engineering from the Indian Institute of Technology, Madras, in 1984 and 1994 respectively. Dr. Asari had been working as an Assistant Professor in Electronics and Communications at TKM College of Engineering, University of Kerala, India. In 1996, he joined the National University of Singapore as a Research Fellow and led the research team for the development of a vision-guided micro-robotic endoscopy system. He joined the School of Computer Engineering, Nanyang Technological University, Singapore in 1998 and led the computer vision and image processing-related research activities in the Center for High Performance Embedded Systems at NTU. Dr. Asari joined the Department of Electrical and Computer Engineering at ODU as an Associate Professor in August 2000 and became a Full Professor in May 2007. He has supervised 14 Ph.D. dissertations and 29 MS theses in Electrical and Computer Engineering at ODU and UD. He has so far published 355 papers including 71 peer reviewed journal papers. Dr. Asari received the Outstanding Teacher Award from the Department of Electrical and Computer Engineering in April 2002 and the Excellence in Teaching Award from the Frank Batten College of Engineering and Technology in April 2004. He also received the Outstanding Researcher Award from the ECE Department and the Excellence in Research Award from the College of Engineering and Technology, both in April 2006. Dr. Asari was awarded two United States patents in 2008 with his former graduate students. He has been a Senior Member of the IEEE since 2001 and is a Senior Member of the Society of Photo-Optical Instrumentation Engineers (SPIE). He is a Member of the IEEE Computational Intelligence Society (CIS). Dr. Asari had been a member of the IEEE CIS Intelligent Systems Applications Technical Committee, IEEE Computer Society, IEEE Circuits and Systems Society, Association for Computing Machinery (ACM), and American Society for Engineering Education (ASEE).

Background Subtraction: Theory and Practice

Ahmed Elgammal

Abstract Background subtraction is a widely-used concept utilized to detect moving objects in videos taken from a static camera. In the last two decades, several algorithms have been developed for background subtraction and were used in various important applications such as visual surveillance, sports video analysis, motion capture, etc. Various statistical approaches have been proposed to model scene backgrounds. In this chapter we review the concept and the practice in background subtraction. We discuss several basic statistical background subtraction models, including parametric Gaussian models and nonparametric models. We discuss the issue of shadow suppression, which is essential for human motion analysis applications. We also discuss approaches and tradeoffs for background maintenance. We also point out many of the recent developments in the background subtraction paradigm.

1 Introduction

In visual surveillance applications, stationary or pan-tilt-zoom (PTZ) cameras are used to monitor activities at outdoor or indoor sites. Since the cameras are stationary, the detection of moving objects can be achieved by comparing each new frame with a representation of the scene background. This process is called *background subtraction* and the scene representation is called the *background model*. The scene here is assumed to be stationary or quasi-stationary.

A. Elgammal (✉)
Rutgers University, New Brunswick, NJ, USA
e-mail: elgammal@cs.rutgers.edu

Augment Vis Real (2014) 6: 1–21
DOI: 10.1007/8612_2012_1

Published Online: 12 September 2012

The concept of background modeling is rooted in photography since the nineteenth century, where it was shown that a film could be exposed for an extended period of time to capture the scene background without moving objects [16]. The use of background subtraction to detect moving objects is deeply rooted in image analysis and emanated from the concept of change detection, a process in which two images of the same scene taken at different time instances are compared; for example in Landsat Imagery, e.g. [10, 29].

Typically, the background subtraction process forms the first stage in automated visual surveillance systems, as well as other applications such as motion capture, sport analysis, etc. Results from background subtraction are used for further processing, such as tracking targets and understanding events. One main advantage of target detection using background subtraction is that the outcome is an accurate segmentation of the foreground regions from the scene background. For human subjects, the process gives accurate silhouettes of the human body which can be further used for tracking, fitting body limbs, pose and posture estimation, etc. This is in contrast to classifier-based, object-based detectors that mainly decides whether a bounding box or a region in the image contains the object of interest or not, such as pedestrian detectors, e.g. [6].

The concept of background subtraction has been widely used since the early human motion analysis systems such as Pfinder [59], W4 [22], etc. Efficient and more sophisticated background subtraction algorithms that can address challenging situations have been developed since then. The success of these algorithms lead to the growth of the automated visual surveillance industry as well as many commercial applications, such as sports monitoring. Unlike earlier background subtraction algorithms, the cameras and the scenes are assumed to be stationary, many approaches have been proposed to overcome these limitations, as in dealing with quasi-stationary scenes and moving cameras. We will discuss such approaches later in this chapter.

The organization of this chapter is as follows. Section 2 discusses some of the challenges in building a background model for detection. Section 3 discusses some of the basic and widely-used background modeling techniques. Section 4 discusses how to deal with color information to avoid detecting shadows. Section 5 discusses the tradeoffs and challenges in updating background models. Section 6 discusses some background models that can deal with moving cameras.

2 Challenges in Scene Modeling

In any indoor or outdoor scene there are changes that occur over time to the scene background. It is important for any background model to be able to tolerate these changes, either by being invariant to them or by adapting to them. These changes can be local, affecting only parts of the background, or global, affecting the entire background. The study of these changes is essential to understand the motivations behind different background subtraction techniques. Toyama et al. [57] identified a list of ten challenges that a background model has to overcome, and denoted them by: *Moved objects, Time of day, Light switch, Waving trees, Camouflage, Bootstrapping, Foreground aperture, Sleeping person, Walking person, Shadows*. Elgammal et al. [14] classifies the possible changes in a scene background according to their source:

Illumination changes:

- Gradual change in illumination as might occur in outdoor scenes due to the change in the relative location of the sun during the day.
- Sudden change in illumination as might occur in an indoor environment by switching the lights on or off, or in an outdoor environment, e.g., a change between cloudy and sunny conditions.
- Shadows cast on the background by objects in the background itself (e.g., buildings and trees) or by moving foreground objects, i.e., moving shadows.

Motion changes:

- Global image motion due to small camera displacements. Despite the assumption that cameras are stationary, small camera displacements are common in outdoor situations due to wind load or other sources of motion which causes global motion in the images.
- Motion in parts of the background. For example, tree branches moving with the wind, or rippling water.

Structural changes:

These are changes introduced to the background, including any change in the geometry or the appearance of the background of the scene introduced by targets. Such changes typically occur when something relatively permanent is introduced into the scene background. For example, if somebody moves (introduces) something from (to) the background, or if a car is parked in the scene or moves out of the scene, or if a person stays stationary in the scene for an extended period, etc. Toyama et al. [57] denoted these situations by "Moved Objects," "sleeping person," and "walking person" scenarios.

A central issue in building a representation for the scene background is what features to use for this representation or, in other words, what to model in the background. In the literature a variety of features have been used for background modeling including pixel based features (pixel intensity, edges, disparity) and region based features (e.g., image blocks). The choice of the features affects how the background model will tolerate the changes in the scene and the granularity of the detected foreground objects.

Another fundamental issue in building a background representation is the choice of the statistical model that explains the observation at a given pixel or region in the scene. The choice of the proper model depends on the type of changes expected in the scene background. Such a choice highly affects the accuracy of the detection. Section 3 discusses some of the statistical models that are widely used in background modeling context. Beyond choosing the features and the statistical model, maintaining the background representation is another challenging issue that we will discuss in Sect. 5.

3 Statistical Scene Modeling

In this section we will discuss some of the existing and widely-used statistical background modeling approaches. For each model we will discuss how the model is initialized and how it is maintained. For simplicity of the discussion we will use pixel intensity as the observation. Instead, color or any other features can be used. At the pixel level, the process of background subtraction can be formulated as followed: given the intensity observed at a pixel at time t, denoted by x_t, we need to classify that pixel to either the background $\mathscr{B}$ or foreground classes $\mathscr{F}$. This is a two-class classification problem. However, since the intensity of a foreground pixel can arbitrary take any value, unless some further information about the foreground is available, we can just assume that the foreground distribution is uniform. Therefore, the problem reduces to a one-class classification problem. We therefore need to model the likelihood of the observation given the background class, i.e., $p(x_t|\mathscr{B})$, which can be achieved if a history of background observations are available at that pixel. If the previous observations are not purely coming from the background, i.e., foreground objects are present in the scene, the problem becomes more challenging.

3.1 Parametric Background Models

3.1.1 Single Gaussian Background Model

Pixel intensity is the most commonly-used feature in background modeling. In a completely static scene, a simple noise model that can be used is an independent stationary additive Gaussian noise model [15]. According to that model, the noise distribution at a given pixel is a zero mean Gaussian distribution $\mathcal{N}(0, \sigma^2)$. It follows that the observed intensity at that pixel is a random variable with a Gaussian distribution, i.e. $P(x_t|\mathscr{B}) \approx \mathcal{N}(\mu, \sigma^2)$. This Gaussian distribution model for the intensity value of a pixel is the underlying model for many background subtraction techniques and widely known as a single Gaussian background model. For the case of color images, a multivariate Gaussian is used. Typically, the color channels are assumed to be independent, which reduces a multivariate Gaussian to a product of single Gaussians, one for each color channel. More discussion about dealing with color will be presented in Sect. 4.

Estimating the parameters for this model, i.e., learning the background model, reduces to estimating the sample mean and variance from history pixel observations. The background subtraction process in this case is a classifier that decides whether a new observation at that pixel comes form the learned background distribution. Assuming the foreground distribution is uniform, this amounts to putting a threshold on the tail of the Gaussian likelihood, i.e., the classification rule reduces to marking a pixel as foreground if

$$\|x_t - \hat{\mu}\| > \text{threshold},$$

where the threshold is typically set to $k\hat{\sigma}$. Here $\hat{\mu}$ and $\hat{\sigma}$ are the estimated mean and standard deviation and k is a free parameter. The standard deviation σ can be assumed to be the same for all pixels. So, literally, this simple model reduces to subtracting a background image B from the each new frame I_t and checking the difference against a threshold. In such a case, the background image B is the mean of the previous background frames.

This basic single Gaussian model can be made adaptive to slow changes in the scene (for example, gradual illumination changes) by recursively updating the mean with each new frame to maintain a background image

$$B_t = \frac{t-1}{t} B_{t-1} + \frac{1}{t} I_t,$$

where $t \geq 1$. Obviously this update mechanism does not forget the history and, therefore, the effect of new images on the model tends to zero. This is not suitable when the goal is to adapt the model to illumination changes. Instead, the mean and variance can be computed over a sliding window of time. However, a more practical and efficient solution is to recursively update the model via temporal blending, also known as exponential forgetting, i.e.

$$B_t = \alpha I_t + (1 - \alpha) B_{t-1}. \tag{1}$$

Here, B_t denotes the background image computed up to frame t. The parameter α controls the speed of forgetting old background information. This update equation is a low-pass filter with a gain factor α that effectively separates the slow temporal process (background) from the fast process (moving objects). Notice that the computed background image is no longer the sample mean over the history but captures the central tendency over time [18].

This basic adaptive model seems to be a direct extension to earlier work on change detection between two images. One of the earliest papers that suggested this model, with a full justification of the background and foreground distributions, is the paper by Donohoe et al. [8]. Karmann et al. [32, 33] used a similar recursive update model without explicit assumption about the background process. Koller et al. [36] used a similar model for traffic monitoring. A similar model was also used in early people tracking systems such as the Pfinder [59].

3.1.2 Mixture Gaussian Background Model

Typically, in outdoor environments with moving trees and bushes, the scene background is not completely static. For example, one pixel can be the image of the sky in one frame, a tree leaf in another frame, a tree branch in a third frame, and some mixture in subsequent frames. In each situation, the pixel will have a different intensity (color), so a single Gaussian assumption for the probability density function of the pixel intensity will not hold. Instead, a generalization based on a Mixture of Gaussians (MoG) has been proposed in [16, 19, 51, 54] to model such variations. This model was introduced by Friedman and Russell [16], where a mixture of three Gaussian distributions was used to model the pixel value for traffic surveillance applications. The pixel intensity was modeled as a weighted mixture of three Gaussian distributions corresponding to road, shadow, and vehicle distribution. Fitting a mixture of Gaussian (MoG) model can be achieved using the Expectation Maximization (EM) algorithm [7]. However, this is impractical for a realtime background subtraction application. An incremental EM algorithm [43] was used to learn and update the parameters of the model.

Stauffer and Grimson [19, 54] proposed a generalization to the previous approach. The intensity of a pixel is modeled by a mixture of K Gaussian distributions (K is a small number from 3 to 5). The mixture is weighted by the frequency with which each of the Gaussians explains the background. The likelihood that a certain pixel has intensity x_t at time t is estimated as

$$p(x_t|\mathscr{B}) = \sum_{i=1}^{K} w_{i,t} G(x_t; \mu_{i,t}, \Sigma_{i,t}), \tag{2}$$

where $w_{i,t}$, $\mu_{i,t}$, and $\Sigma_{i,t} = \sigma_{i,t}\mathbf{I}$ are the weight, mean, and covariance for the ith Gaussian mixture component at time t respectively.

The parameters of the distributions are updated recursively using online K-means approximation. The mixture is weighted by the frequency with which each of the Gaussians explains the background, i.e., a new pixel value is checked against the existing K Gaussians, and when a match is found, the weight for that distribution is updated as follows

$$w_{i,t} = (1 - \alpha) w_{i,t-1} + \alpha M(i, t),$$

where $M(i,t)$ is an indicator variable, which is 1 if the ith component is matched, 0 otherwise. The parameter of the matched distributions are updated as follows

$$\mu_t = (1 - \rho)\mu_{t-1} + \rho x_t,$$

$$\sigma_t^2 = (1 - \rho)\sigma_{t-1}^2 + \rho (x_t - \mu_t)^T (x_t - \mu_t).$$

The parameters α and ρ are two learning rates. The K distributions are ordered based on w_j/σ_j^2 and the first B distributions are used as a model of the background of the scene where B is estimated as

$$B = \arg\min_b \left(\sum_{j=1}^{b} w_j > T \right). \tag{3}$$

The threshold T is the fraction of the total weight given to the background model. Background subtraction is performed by marking any pixel that is more that 2.5 standard deviations away from any of the B distributions as a foreground pixel.

The MoG background model was shown to perform very well in indoor and outdoor situations. Many variations has been suggested to Stauffer and Grimson's model [54], e.g. [23, 30, 41], etc. The model also was used with different feature spaces and/or with a subspace representations. Gao et al. [18] studied the statistical error characteristic of MoG background models.

3.2 Non-Parametric Background Models

Typically in outdoor scenes, there are wide range of variations, which can be very fast. Outdoor scenes usually contains dynamic areas such as waving trees and bushes, rippling water, or ocean waves. Such variations are part of the scene background. Modeling such dynamic areas requires a more flexible representation of the background probability distribution at each pixel. This motivates the use of non-parametric density estimator for background modeling [12].

A particular non-parametric technique that estimates the underlying density and is quite general is the kernel density estimation (KDE) technique [9, 48]. Given a sample $S = \{x_i\}_{i=1..N}$ from a distribution with density function $p(x)$, an estimate $\hat{p}(x)$ of the density at x can be calculated using

$$\hat{p}(x) = \frac{1}{N} \sum_{i=1}^{N} K_\sigma(x - x_i), \tag{4}$$

where K_σ is a kernel function (sometimes called a "window" function) with a bandwidth (scale) σ such that $K_\sigma(t) = \frac{1}{\sigma} K(\frac{t}{\sigma})$. The kernel function K should satisfy $K(t) \geq 0$ and $\int K(t)dt = 1$. Kernel density estimators asymptotically converge to any density function with sufficient samples [9, 48]. In fact, all other nonparametric density estimation methods, e.g., histograms, can be shown to be asymptotically kernel methods [48]. This property makes these techniques quite general and applicable to many vision problems where the underlying density is not known [3, 11]. We can avoid having to store the complete data set by weighting a subset of the samples as

$$\hat{p}(x) = \sum_{x_i \in B} \alpha_i K_\sigma(x - x_i),$$

where α_i are weighting coefficients that sum up to one and B is a sample subset. A good discussion of KDE techniques can be found in [48].

Elgammal et al. [12] introduced a background modeling approach based on kernel density estimation. Let $x_1, x_2, ..., x_N$ be a sample of intensity values for a pixel. Given this sample, we can obtain an estimate of the probability density function of the pixel intensity at any intensity value using kernel density estimation using Eq. 4. This estimate can be generalized to use color features or other high-dimensional features by using kernel products as

$$p(x_t|\mathscr{B}) = \frac{1}{N}\sum_{i=1}^{N}\prod_{j=1}^{d} K_{\sigma_j}(x_t^j - x_i^j), \tag{5}$$

where x_t^j is the jth dimension of the color features at time t and K_{σ_j} is a kernel function with bandwidth σ_j in the jth color space dimension.

A variety of kernel functions with different properties have been used in the literature of non-parametric estimation. Typically, kernel functions are symmetric and unimodal functions, which fall off to zero rapidly away from the center, i.e., the kernel function should have finite local support and points beyond certain window will have no contribution. The Gaussian function is typically used as a kernel for its continuity, differentiability, and locality properties; although it violates the finite support criterion [11]. Note that choosing the Gaussian as a kernel function is different from fitting the distribution to a Gaussian model. Here, the Gaussian is only used as a function to weigh the data points. Unlike parametric fitting of a mixture of Gaussians, kernel density estimation is a more general approach that does not assume any specific shape for the density function.

Using this probability estimate, the pixel is considered a foreground pixel if $p(x_t|\mathscr{B}) < th$, where the threshold th is a global threshold over the whole image that can be adjusted to achieve a desired percentage of false positives. Practically, the probability estimation in Eq. 5 can be calculated in a very efficient way using precalculated lookup tables for the kernel function values given the intensity value difference, $(x_t - x_i)$, and the kernel function bandwidth. Moreover, a partial evaluation of the sum in Eq. 5 is usually sufficient to surpass the threshold at most image pixels, since most of the image is typically from the background. This allows a realtime implementation of the approach.

The estimate in Eq. 4 is based on the most recent N samples used in the computation. Therefore, adaptation of the model can be achieved simply by adding new samples and ignoring older samples [12], i.e., using a sliding window over time.

This nonparametric technique for background subtraction was introduced in [12] and has been tested for a wide variety of challenging background subtraction problems in a variety of setups and was found to be robust and adaptive. We refer the reader to [12] for details about the approach, such as details about model adaptation and false detection suppression. Figure 1 shows two detection results for targets in a wooded area, where tree branches move heavily and the target is highly occluded. Figure 2—top shows the detection results using an omni-directional camera. The targets are camouflaged and walking through the woods. Figure 2—bottom shows the detection result for a rainy day, where the background model adapts to account for different rain and lighting conditions.

Fig. 1 Example background subtraction detection results: *left* original frames, *right* detection results

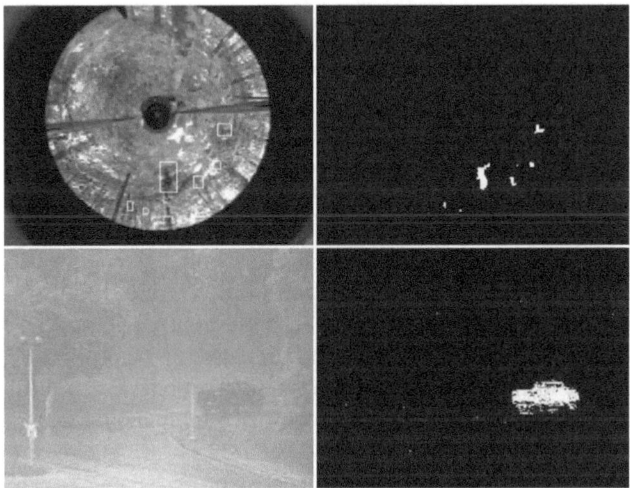

Fig. 2 *Top* detection of camouflaged targets from an omni-directional camera. *Bottom* detection result for a rainy day

One major issue that needs to be addressed when using kernel density estimation technique is the choice of suitable kernel bandwidth (scale). Theoretically, as the number of samples increases, the bandwidth should decrease. Practically, since only a finite number of samples are used, and the computation must be performed in real time, the choice of a suitable bandwidth is essential. Too small bandwidth will lead to a ragged density estimate, while a too wide bandwidth will lead to an over-smoothed density estimate [9, 11]. Since the expected variations in pixel intensity over time are different from one location to another in the image, a different kernel bandwidth is used for each pixel. Also, a different kernel bandwidth is used for each color channel. In [12] a procedure was proposed for estimating the kernel bandwidth for each pixel as a function of the median of absolute differences between consecutive frames. In [42] an adaptive approach for estimation of kernel bandwidth was proposed. Parag et al. [44] proposed an approach using boosting to evaluate different kernel bandwidth choices for bandwidth selection.

KDE-Background Practice and Other Nonparametric Models:

One of the drawbacks of the KDE background model is the requirement to store a large number of history samples for each pixel. In KDE literature, many approaches were proposed to avoid storing a large number of samples. Within the context of background modeling, Piccardi and Jan [46] proposed an efficient mean shift approach to estimate the modes of a pixel's history PDF, then a few number of Gaussians was used to model the PDF. Mean shift is a non-parametric iterative mode-seeking procedure [2, 4, 17]. With the same goal of reducing memory requirement, Han et al. [21] proposed a sequential kernel density estimation approach, where variable bandwidth mean shift was used to detect the density modes. Unlike mixture of Gaussian methods where the number of Gaussians is fixed, techniques such as [21, 46] can adaptively estimate a variable number of modes to represent the density, therefore, keeping the flexibility of a nonparametric model while achieving the efficiency of a parametric model.

Efficient implementation of KDE can be achieved through building look-up tables for the kernel function values, which facilitates realtime performance. Fast Gauss Transform has been proposed for efficient computation of KDE [13], however, the Fast Gauss Transform is only justifiable with a large number of samples required for the density estimation, as well as the need for estimation at many pixels in batches. For example, Fast Gauss implementation was effectively used in a layered background representation [45].

Many variations have been suggested to the basic nonparametric KDE background model. In practice, non-parametric KDE has been used at the pixel level as well as at the region level, or in a domain-range representation to model a scene background. For example, in [45] a layered representation was used to model the scene background where the distribution of each layer is modeled using KDE. Such layered representation facilitates detecting the foreground under static or dynamic background and in the presence of nominal camera motion. In [49] KDE was used in a joint domain-range representation of image pixel (r, g, b, x, y), which exploits the spatial correlation between neighboring pixels. Parag et al. [44] proposed an approach for feature selection for the KDE framework where boosting based ensemble learning was used to combine different features. The approach also can be used to evaluate different choices for the kernel bandwidth. Recently, Sheikh et al. [50] used a KDE approach in a joint domain-range representation within a foreground/background segmentation framework from freely moving camera as will be discussed in Sect. 6.

3.3 Other Statistical Models

The statistical models for background subtraction that are described in this section are the basis for many other algorithms in the literature. In [57], linear prediction using the Wiener filter is used to predict pixel intensity given a recent history of values. The prediction coefficients are recomputed at each frame from the sample covariance to achieve adaptivity. Linear prediction using the Kalman Filter was also used in [32, 33, 36]; such models assume a single Gaussian distribution of the pixel process.

Another approach used to model a wide range of variations in the pixel intensity is to represent these variations as discrete states corresponding to modes of the environment, e.g., lights on/off, cloudy/sunny. Hidden Markov Models (HMM) have been used for this purpose in [47, 55]. In [47], a three-state HMM has been used to model the intensity of a pixel for traffic monitoring application, where the three states correspond to the background, shadow, and foreground. The use of HMMs imposes a temporal-continuity constraint on the pixel intensity, i.e., if the pixel is detected as a part of the foreground, then it is expected to remain part of the foreground for a period of time before switching back to be part of the background. In [55], the topology of the HMM representing global image intensity is learned while training the background model. At each global intensity state, the pixel intensity is modeled using a single Gaussian. It was shown that the model is able to learn simple scenarios like switching the lights on or off.

Background subtraction techniques can successfully deal with quasi-moving background, e.g. scenes with dynamic textures. The non-parametric model using Kernel Density Estimation (KDE), described above, has very good performance in scenes with dynamic backgrounds, such as outdoor scenes with trees in the background. Several approaches were developed to address such dynamic scenes. In [52] an Auto Regressive Moving Average (ARMA) model was proposed for modeling dynamic textures. ARMA is a first-order linear-prediction model. In [63] an ARMA model was used for background modeling of scenes with dynamic texture, where a robust Kalman filter were used, to update the model. In [42] a combination of optical flow and appearance features were used, within an adaptive kernel density estimation framework, to deal with dynamic scenes.

In [38] a biologically-inspired non-parametric background subtraction approach was proposed, where a self-organizing artificial neural network model was used to model the pixel process. Each pixel is modeled with a sample arranged in a shared 2D grid of nodes, where each node is represented with a weight vector with the same dimensionality as the input observation. An incoming pixel observation is mapped to the node whose weights are most similar to the input where a threshold function is used to decide background/foreground. The weights of each node are updated at each new frame using a recursive filter similar to Eq. 1. An interesting feature of this approach is that the shared 2D grid of nodes allows the spatial relationships between pixels to be taken into account at both the detection and update phases.

3.4 Features for Background Modeling

Intensity has been the most commonly-used feature for modeling the background. Alternatively, edge features have also been used to model the background. The use of edge features to model the background is motivated by the desire to have a representation of the scene background that is invariant to illumination changes, as discussed in Sect. 4. In [60] foreground edges are detected by comparing the edges in each new frame with an edge map of the background, which is called the background "primal sketch." The major drawback of using edge features to model the background is that it would only be possible to detect edges of foreground objects instead of the densely-connected regions that result from pixel intensity-based approaches. Fusion of intensity and edge information was used in [27, 28, 40, 62]. Among many other feature studied, optical flow was used in [42] to help capture background dynamics. A general framework for feature selection based on boosting for background modeling was proposed in [44].

Besides pixel-based approaches, block-based approaches have also been used for modeling the background. Block matching has been extensively used for change detection between consecutive frames. In [26] each image block is fit to a second-order bivariate polynomial and the remaining variations are assumed to be noise. A statistical likelihood test is then used to detect blocks with significant change. In [39] each block was represented with its median template over the background learning period and its block standard deviation. Subsequently, at each new frame, each block is correlated with its corresponding template and blocks with too much deviation relative to the measured standard deviation are considered to be foreground. The major drawback with block-based approaches is that the detection unit is a whole image block and therefore they are only suitable for coarse detection.

4 Moving Shadow Suppression

A background subtraction process on gray scale images, or on color images without carefully selecting the color space, is bound to detect the shadows of moving objects along with the objects themselves. While shadows of static objects can typically be adapted in the background process, shadows casted by moving object, i.e., dynamic shadows, constitute a sever challenge for foreground segmentation. Since the goal of background subtraction is to obtain accurate segmentation of moving foreground regions for further processing, it is highly desirable to detect such foreground regions without casted shadow attached to them. This is particularly important for human motion analysis, since shadows attached to silhouettes would cause problems in fitting body limbs and estimating body poses; consider the example shown in Fig. 3. Therefore, extensive research has addressed the detection/supression of moving (dynamic) shadows.

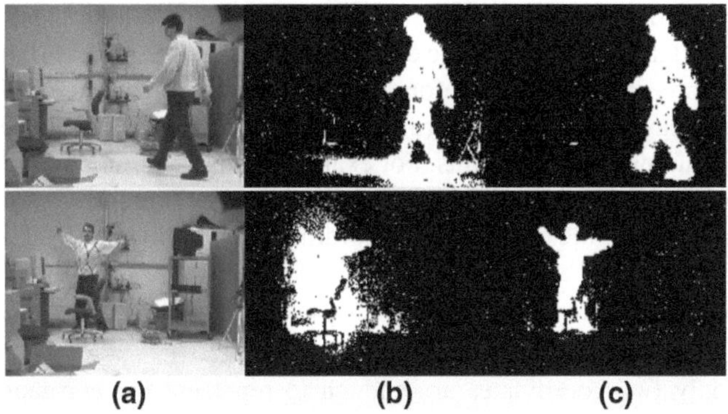

Fig. 3 **a** Original frames, **b** detection using (R, G, B) color space, **c** detection using chromaticity coordinates (r, g) and the lightness variable s

Avoiding the detection of shadows or suppressing the detected shadows can be achieved in color sequences by understanding how shadows affect color images. This is also useful to achieve a background model that is invariant to illumination changes. Cast shadows has a dark part (umbra) where a light source is totally occluded, and a soft transitional part (penumbra) where light is partially occluded [53]. In visual surveillance scenarios, the penumbra shadows are common, since diffused and indirect light is common in indoor and outdoor scenes. Penumbra shadows can be characterized by a low value of intensity while preserving the chromaticity of the background, i.e. achromatic shadows. Most research on detecting shadows have focused on achromatic shadows [5, 12, 25].

Let us consider the RGB color space, which is a typical output of a color camera. The brightness of a pixel is a linear combination or the RGB channels, here denoted by I

$$I = w_r R + w_g G + w_b B.$$

When an object casts a shadow on a pixel, less light reaches it and it seems darker. Therefore, a shadow casted on a pixel can be characterized by a change of in brightness of that pixel such that

$$\tilde{I} = \alpha I$$

where $\tilde{I}$ is the pixel's new brightness. Similar effect happens under certain changes in illumination, e.g., turning on/off the lights. Here $\alpha < 1$ for the case of shadow, which means the pixel is darker under shadow, while $\alpha > 1$ for the case of highlights, the pixel seems brighter. A change in the brightness of a pixel will affect all the three color channels R, G, and B. Therefore any background model based on the RGB space, and of course gray scale imagery, is bound to detect moving shadows as foreground regions.

So, which color spaces are invariant or less sensitive to shadows and highlights? For simplicity, let us assume that the effect of the change in a pixel brightness is the same in the three channels. Therefore, the observed colors are αR, αG, αB. Any chromaticity measure of a pixel where the effect of the α factor is cancelled, is in fact invariant to shadows and highlights. For example, in [12] chromaticity coordinates based on normalized RGB were used for modeling the background. Given three color variables, R, G and B, the chromaticity coordinates are defined as [37]

$$r = \frac{R}{R+G+B}, g = \frac{G}{R+G+B}, b = \frac{B}{R+G+B} \tag{6}$$

Obviously only two coordinates are enough to represent the chromaticity, since $r+g+b=1$. The above equation describes a central projection to the plane $R+G+B=1$.[1] It can be easily seen that the chromaticity variables r, g, b are invariant to shadows and highlights (according to our assumption) since the α factor does not have an effect on them. Figure 3 shows the results of detection using both (R, G, B) space and (r, g) space. The figure shows that using the chromaticity coordinates allows detection of the target without detecting its shadow.

Some other color spaces also have chromaticity variables that are invariant to shadows and highlights in the same way. For example, the reader can verify that the Hue and Saturation variables in the HSV color space are invariant to the α factor and thus insensitive to shadows and highlights, while the Value variable, which represents the brightness, is variant to them. Therefore, the HSV color space has been used in some background subtraction algorithms that suppress shadows, e.g. [5]. Similarly, HSL and CIE-xy spaces have the same property. On the other hand, color spaces such as YUV, YIQ, YCbCr are not invariant to shadows and highlights since they are just linear transformations from the RGB space.

Although using chromaticity coordinates helps in the suppression of shadows, they have the disadvantage of losing lightness information. Lightness is related to the differences in whiteness, blackness, and grayness between different objects [21]. For example, consider the case, where the target wears a white shirt and walks against a gray background. In this case, there is no color information. Since both white and gray have the same chromaticity coordinates, the target will not be detected using only chromaticity variables. In fact, in the r, g space, the whole gray line (R = G = B) projects to the point (1/3, 1/3) in the space; similarly for CIE xy. Therefore, there is no escape of using a brightness variable! In [12] a third "lightness" variable $s = R+G+B$ was used besides r, g. While the chromaticity variable r, g are not expected to change under shadow, s is expected to change

[1] This is analogous to the transformation used to obtain CIE xy chromaticity space from CIE XYZ color space. The CIE XYZ color space is a linear transformation to the RGB space [1]. The chromaticity space defined by the variable r,g is therefore analogous to the CIE xy chromaticity space.

within limits, which corresponds to the expected shadows and highlights in the scene.

Most approaches for shadow suppression rely on the above reasoning of separating the chromaticity distortion, from brightness distortion, where each of these distortions are treated differently, e.g. [5, 12, 25, 27, 34]. In [25] both brightness and color distortions are defined using a chromatic cylinder model. By projecting an observed pixel color to the vector defined by that pixel's background value in the RGB color space (chromaticity line), the color distortion is defined as the orthogonal distance, while the projection defines the brightness distortion. Here, a single Gaussian background model is assumed. These two measures were used to classify an observation to either background, foreground, shadows, or highlights. Notice that the orthogonal distance between an observed pixel's RGB color and a chromaticity line is affected by brightness of that pixel, while the distance measured in the r–g space (or xy space) corresponds to the angles between the observed color vector and the chromaticity line, i.e., the r–g space used in [12] is a projection of a chromatic cone. In [27] a chromatic and brightness distortion model is used similar to [25, 34], although using a chromatic cone instead of a chromatic cylinder distortion model.

Another class of algorithms for shadow suppression is the class of approaches that depend on image gradient to model the scene background. The idea is that texture information in the background will be consistent under shadow, so using the image gradient as a feature will be invariant to cast shadows, except at the shadow boundary. These approaches utilize a background edge or gradient model besides the chromaticity model to detect shadows, e.g. [27, 28, 40, 62]. In [27] a multistage approach was proposed to detect chromatic shadows. In the first stage, potential shadow regions are detected by fusing color (using the invariant chromaticity cone model described above) and gradient information. In the second stage, pixels in these regions are classified using different cues, including spatial and temporal analysis of chrominance, brightness, texture distortion, and a measure of diffused sky lighting denoted by "bluish effect." The approach can successfully detect chromatic shadows.

5 Tradeoffs in Background Maintenance

As discussed in Sect. 1 there are different changes that can occur in a scene background, which can be classified to: illumination changes, Motion Changes, and Structural Changes. The goal of background maintenance is to cope with these changes and keep an updated version of the scene background model. In parametric background models, recursive update in the form of Eq. 1 (or some variant of it) is typically used for background maintenance, e.g. [32, 36, 54]. In nonparametric models, the sample of each pixel history is updated continuously in order to achieve adaptability [12, 42]. These recursive updates, along with careful choice of the color space, are typically enough to deal with both the illumination changes and motion changes previously described.

The most challenging cases are where changes are introduced to the background (objects moved in or from the background) denoted here by "Structural Changes." For example, when a vehicle came and parked in the scene. A background process should detect such a vehicle but should also adapt it to the background model in order to be able to detect other targets that might pass in front of it. Similarly if a vehicle that was already part of the scene moved out, a false detection 'hole' will appear in the scene where that vehicle was parked. There are many examples similar to these scenarios. Toyama et al. [57] denoted these situations by "sleeping person" and "walking person" scenarios.

Here we point out two interwound tradeoffs that associate with maintaining any background model.

Background update rate: The speed or the frequency in which a background model gets updated highly influences the performance of the process. In most parametric models, the learning rate α in Eq. 1 controls the speed in which the model adapts to changes. In non-parametric models, the frequency in which new samples are added to the model has the same effect. Fast model update makes the model able to rapidly adapt to scene changes such as fast illumination changes, which leads to high sensitivity in foreground/background classification. However, the model can also adapt to targets in the scene if the update is done blindly in all pixels, or errors occurs in masking out foreground regions. Slow update is a safer way to avoid integrating any transient changes to the model. However, the classifier will lose its sensitivity in case of fast scene changes.

Selective versus blind update: Given a new pixel observation, there are two alternative mechanisms for updating a background model: (1) Selective Update: update the model only if the pixel is classified as a background sample. (2) Blind Update: just update the model regardless of the classification outcome. Selective update is commonly used by masking out foreground-classified pixels from the update, since updating the model with foreground information would lead to increased false negative, e.g., holes in the detected targets. The problem with selective update is that any incorrect detection decision will result in persistent incorrect detection later, which is a deadlock situation, as denoted by Karmann et al. [32]. For example, if a tree branch is displaced and stayed fixed in the new location for a long time, it would be continually detected. This is what leads to the 'Sleeping/Walking person' problems as denoted in [57].

Blind update does not suffer from this deadlock situations, since it does not involve any update decisions; it allows intensity values that do not belong to the background to be added to the model. This might lead to more false negatives as targets erroneously become part of the model. This effect can be reduced if the update rate is slow.

The interwound effects of these two tradeoffs is shown in Table 1. Most background models chose a selective update approach and try to avoid the effects of detection errors by using a slow update rate. However, this is bound to cause deadlocks. In [12] the use of a combination of two models was proposed: a short-term model (selective and fast) and a long-term model (blind and slow). This

Table 1 Tradeoffs in background maintenance

	Fast update	Slow update
Selective update	Highest sensitivity	Less sensitivity
	Adapts to fast illumination changes	
	Bound to deadlocks	Bound to deadlocks
Blind update	Adapts to targets (more false negatives)	Slow adaptation
	No deadlocks	No deadlocks

combination tries to achieve high sensitivity and, at the same time, avoids deadlocks.

Several approaches have been proposed for dealing with specific scenarios with structural changes. The main problem is that dealing with such changes requires a higher level of reasoning about which objects are causing such structural changes (vehicle, person, animal) and what should be done with them, which mostly depends on the application. Such high-level of reasoning is typically beyond the design goal of the background process, which is mainly a low-level process that knows only about pixel appearance.

The idea of using multiple background models was further developed by Kim et al. in [35] to address scene structure changes in an elegant way. In that work, a layered background model was used, where a long-term background model is used, besides several multiple short-term background models that capture temporary changes in the background. An object that comes into the scene and stops is represented by a short-term background (layer). Therefore, if a second object passes in front of the stopped object, it will also be detected and represented as a layer as well. Figure 4 shows an overview of that approach and detection results.

6 Background Subtraction for Wide-Area Surveillance

A fundamental limitation for background subtraction techniques is the assumption of a stationary camera. Several approaches have been suggested to alleviate this constraint and develop background subtraction techniques that can work with moving camera under some motion constraints. Rather than a pixel-level representation, a region-based representation of the scene background can help tolerate some degree of camera motion, e.g. [45]. In particular, the case of a pan-tilt-zoom (PTZ) camera has been addressed because of its importance in surveillance applications. If the camera motion is pure rotation with no translation (or close to zero baseline), camera motion can be modeled by a Homography and image mosaicing approaches can be used to built a background model. Physically panning and tilting are not pure rotations, since the rotation of the camera head is not around the center of projection. Therefore, a small translation is involved in the camera motion producing a parallax effect. There have been several approaches for

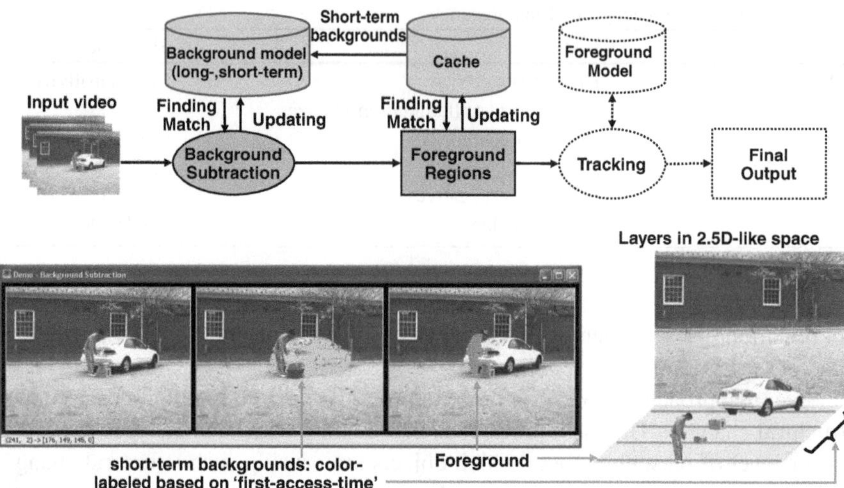

Fig. 4 An overview of Kim et al. approach [35] with short-term background layers: the foreground and the short-term backgrounds can be interpreted in a different temporal order

building a background model from a panning camera based on building an image mosaic and the use of a MoG model, e.g. [41, 51]. Mittal and Huttenlocher [41] used image registration to align each new image to a panoramic background image generated from a MoG Model. The registered image is then used to update the background model. Alternatively, in [58] a representation of the scene background as a finite set of images on a virtual polyhedron is used to construct images of the scene background at any arbitrary pan-tilt-zoom setting.

Recently there have been some interests in Background Subtraction/foreground–background separation from freely-moving cameras, e.g. [24, 31, 50]. There is a huge literature on motion segmentation [61], which exploits motion discontinuity. However, these approaches do not necessarily aim at modeling scene background and segmenting the foreground layers. Fundamentally motion segmentation by itself is not enough to separate the foreground from the background in the case where both of them constitute a rigid or close to rigid motion, e.g., a car parked in the street or a person standing will have the same 3D motion w.r.t. the camera as the rest of the scene. Similarly, depth discontinuity by itself is not enough to distinguish the foreground from the background, since objects of interest can be at a distance from the camera with no significant depth difference than the background.

Most notably, Sheikh et al. [50] used affine factorization to develop a framework for moving camera background subtraction. In this approach, trajectories of sparse image features are segmented using affine factorization [56]. A sparse representation of the background is maintained by estimating a trajectory basis that spans the background subspace. KDE was then used to model the appearance of the background and foreground from the sparse features. A Markov Random Field was used to achieve the final labeling.

References

1. Burger, W., Burge, M.: Digital Image Processing, an Algorithmic Introduction Using Java. Springer, New York (2008)
2. Cheng, Y.: Mean shift, mode seeking, and clustering. IEEE Trans. Pattern Anal. Mach. Intell. **17**(8), 790–799 (1995)
3. Comaniciu, D.: Nonparametric robust methods for computer vision. Ph.D. thesis, Rutgers, The State University of New Jersey (2000)
4. Comaniciu, D., Meer, P.: Mean shift analysis and applications. In: IEEE 7th International Conference on Computer Vision, vol. 2, pp. 1197–1203, 1999
5. Cucchiara, R., Grana, C., Piccardi, M., Prati, A.: Detecting moving objects, ghosts, and shadows in video streams. IEEE Trans. Pattern Anal. Mach. Intell. **25**, 1337–1342 (2003)
6. Dalal, N., Triggs, B.: Histograms of oriented gradients for human detection. In: CVPR, 2005
7. Dempster, A., Laird, N., Rubin, D.: Maximum likelihood from incomplete data via the em algorithm. J. Royal Stat. Soc. **39**, 1–38 (1977)
8. Donohoe, G.W., Hush, D.R., Ahmed, N.: Change detection for target detection and classification in video sequences. In: ICASSP, 1988
9. Duda, R.O., Stork, D.G., Hart, P.E.: Pattern Classification. Wiley, New York (2000)
10. Eghbali, H.J.: K-s test for detecting changes from landsat imagery data. SMC **9**(1), 17–23 (1979)
11. Elgammal, A.: Efficient kernel density estimation for realtime computer vision. Ph.D. thesis, University of Maryland, 2002
12. Elgammal, A., Harwood, D., Davis, L.S.: Nonparametric background model for background subtraction. In: Proceedings of 6th European Conference of Computer Vision, 2000
13. Elgammal, A., Duraiswami, R., Davis, L.S.: Efficient non-parametric adaptive color modeling using fast gauss transform. In: Proceedings of IEEE Conference on Computer Vision and Pattern Recognition, 2001
14. Elgammal, A., Duraiswami, R., Harwood, D., Davis, L.S.: Background and foreground modeling using non-parametric kernel density estimation for visual surveillance. In: Proceedings of the IEEE, 2002
15. Forsyth, D.A., Ponce, J.: Computer Vision a Modern Approach. Prentice Hall, Upper Saddle River (2002)
16. Friedman, N., Russell, S.: Image segmentation in video sequences: a probabilistic approach. In: Uncertainty in Artificial Intelligence, 1997
17. Fukunaga, K., Hostetler, L.D.: The estimation of the gradient of a density function, with application in pattern recognition. IEEE Trans. Inf. Theory **21**, 32–40 (1975)
18. Gao, X., Boult, T.E.: Error analysis of background adaption. In: IEEE Conference on Computer Vision and Pattern Recognition, 2000
19. Grimson, W.E.L., Stauffer, C., Romano, R.: Using adaptive tracking to classify and monitor activities in a site. In: IEEE Conference on Computer Vision and Pattern Recognition, 1998
20. Hall, E.L.: Computer Image Processing and Recognition. Academic Press, New York (1979)
21. Han, B., Comaniciu, D., Davis, L.: Sequential kernel density approximation through mode propagation: applications to background modeling. In: Proceedings of ACCV 2004, 2004
22. Haritaoglu, I., Harwood, D., Davis, L.S.: W4: who? when? where? what? a real time system for detecting and tracking people. In: International Conference on Face and Gesture Recognition, 1998
23. Harville, M.: A framework for high-level feedback to adaptive, per-pixel, mixture-of-gaussian background models. In: ECCV, pp. 543–560, 2002

24. Hayman, E., Eklundh, J.O.: Statistical background subtraction for a mobile observer. In: Proceedings ICCV, pp. 67–74, 2003
25. Horprasert, T., Harwood, D., Davis, L.S.: A statistical approach for real-time robust background subtraction and shadow detection. In: IEEE Frame-Rate Applications Workshop, 1999
26. Hsu, Y.Z., Nagel, H.H., Rekers, G.: New likelihood test methods for change detection in image sequences. Comput. Vis. Image Process. **26**, 73–106 (1984)
27. Huerta, I., Holte, M., Moeslund, T., Gonzalez, J.: Detection and removal of chromatic moving shadows in surveillance scenarios. In: ICCV'09, pp. 1499–1506, 2009
28. Jabri, S., Duric, Z., Wechsler, H., Rosenfeld, A.: Detection and location of people in video images using adaptive fusion of color and edge information. In: International Conference of Pattern Recognition, 2000
29. Jain, R.C., Nagel, H.H.: On the analysis of accumulative difference pictures from image sequences of real world scenes. PAMI **1**(2), 206–213 (1979)
30. Javed, O., Shafique, K., Shah, M.: A hierarchical approach to robust background subtraction using color and gradient information. In: IEEE Workshop on Motion and Video Computing, pp. 22–27, 2002
31. Jin, Y.X., Tao, L.M., Di, H., Rao, N.I., Xu, G.Y.: Background modeling from a free-moving camera by multi-layer homography algorithm. In: ICIP, pp. 1572–1575, 2008
32. Karmann, K.-P., von Brandt, A.: Moving object recognition using and adaptive background memory. In: Time-Varying Image Processing and Moving Object Recognition. Elsevier Science Publishers B.V., Amsterdam (1990)
33. Karmann, K.-P., Brandt, A.V., Gerl, R.: Moving object segmentation based on adabtive reference images. In: Signal Processing V: Theories and Application. Elsevier Science Publishers B.V., Amsterdam (1990)
34. Kim, K., Chalidabhongse, T.H., Harwood, D., Davis, L.: Background modeling and subtraction by codebook construction. In: International Conference on Image Processing, pp. 3061–3064, 2004
35. Kim, K., Harwood, D., Davis, L.S.: Background updating for visual surveillance. In: Proceedings of the International Symposium on Visual Computing, pp. 1–337, 2005
36. Koller, D., Weber, J., Huang, T., Malik, J., Ogasawara, G., Rao, B., Russell, S.: Towards robust automatic traffic scene analyis in real-time. In: International Conference of Pattern Recognition, 1994
37. Levine, M.D.: Vision in Man and Machine. McGraw-Hill Book Company, New York (1985)
38. Maddalena, L., Petrosino, A.: A self-organizing approach to background subtraction for visual surveillance applications. IEEE Trans. Image Process. **17**(7), 1168–1177 (2008)
39. Matsuyama, T., Ohya, T., Habe, H.: Background subtraction for nonstationary scenes. In: 4th Asian Conference on Computer Vision, 2000
40. Mckenna, S.J., Sumer, J., Zoran, D., Harry, W., Azriel, R.: Tracking groups of people. Comput. Vis. Image Underst. **80**, 42–56 (2000)
41. Mittal, A., Huttenlocher, D.: Scene modeling for wide area surveillance and image synthesis. In: CVPR, 2000
42. Mittal, A., Paragios, N.: Motion-based background subtraction using adaptive kernel density estimation. In: CVPR, pp. 302–309, 2004
43. Neal, R.M., Hinton, G.E.: A new view of the em algorithm that justifies incremental and other variants. In: Learning in Graphical Models, pp. 355–368. Kluwer Academic Publishers, Dordrecht (1993)
44. Parag, T., Elgammal, A., Mittal, A.: A framework for feature selection for background subtraction. In: Proceedings of IEEE Computer Society Conference on Computer Vision and Pattern Recognition, CVPR06, June 2006
45. Patwardhan, K., Sapiro, G., Morellas, V.: Robust foreground detection in video using pixel layers. IEEE Trans. Pattern Anal. Mach. Intell. **30**, 746–751 (2008)
46. Piccardi M., Jan T, (2004) Mean-shift background image modelling. ICIP **5**, 3399–3402
47. Rittscher, J., Kato, J., Joga, S., Blake, A.: A probabilistic background model for tracking. In: 6th European Conference on Computer Vision, 2000

48. Scott, D.W.: Mulivariate Density Estimation. Wiley-Interscience, Hoboken (1992)
49. Sheikh, Y., Shah, M.: Bayesian modeling of dynamic scenes for object detection. PAMI **27**, 1778–1792 (2005)
50. Sheikh, Y., Javed, O., Kanade, T.: Background subtraction for freely moving cameras. In: ICCV, pp. 1219–1225, 2009
51. Simon, R., Andrew, B.: Statistical mosaics for tracking. Image Vis. Comput. **14**(8), 549–564 (1996)
52. Soatto, S., Doretto, G., Wu, Y.: Dynamic textures. In: Proceedings of the International Conference on Computer Vision, vol. 2, pp. 439–446, 2001
53. Stauder, J., Mech, R., Ostermann, J.: Detection of moving cast shadows for object segmentation. IEEE Trans. Multimedia **1**, 65–76 (1999)
54. Stauffer, C., Grimson, W.E.L.: Adaptive background mixture models for real-time tracking. In: IEEE Conference on Computer Vision and Pattern Recognition, 1999
55. Stenger, B., Ramesh, V., Paragios, N., Coetzee, F., Bouhman, J.: Topology free hidden markov models: application to background modeling. In: IEEE International Conference on Computer Vision, 2001
56. Tomasi, C.: Shape and motion from image streams under orthography: a factorization method. Int. J. Comput. Vis. **9**, 137–154 (1992)
57. Toyama, K., Krumm, J., Brumitt, B., Meyers, B.: Wallflower: principles and practice of background maintenance. In: IEEE International Conference on Computer Vision, 1999
58. Wada, T., Matsuyama, T.: Appearance sphere: background model for pan-tilt-zoom camera. In: 13th International Conference on Pattern Recognition, 1996
59. Wern, C.R., Azarbayejani, A., Darrell, T., Pentland, A.P.: Pfinder: real-time tracking of human body. In: IEEE Transaction on Pattern Analysis and Machine Intelligence, 1997
60. Yang, Y.-.H., Levine, M.D.: The background primal sketch: an approach for tracking moving objects. Mach. Vis. Appl. **5**, 17–34 (1992)
61. Zappella, L., Lladó, X., Salvi, J.: Motion segmentation: a review. In: Proceeding of the 2008 Conference on Artificial Intelligence Research and Development, pp. 398–407. IOS Press, Amsterdam (2008)
62. Zhang, W., Fang, X.Z., Yang, X.: Moving cast shadows detection based on ratio edge. In: International Conference on Pattern Recognition (ICPR), 2006
63. Zhong, J., Sclaroff, S.: Segmenting foreground objects from a dynamic textured background via a robust kalman filter. In: ICCV '03: Proceedings of the Ninth IEEE International Conference on Computer Vision, p. 44. IEEE Computer Society, Washington (2003)

Moving Cast Shadows Detection Methods for Video Surveillance Applications

Ariel Amato, Ivan Huerta, Mikhail G. Mozerov, F. Xavier Roca and Jordi Gonzàlez

Abstract Moving cast shadows are a major concern in today's performance from broad range of many vision-based surveillance applications because they highly difficult the object classification task. Several shadow detection methods have been reported in the literature during the last years. They are mainly divided into two domains. One usually works with static images, whereas the second one uses image sequences, namely video content. In spite of the fact that both cases can be analogously analyzed, there is a difference in the application field. The first case, shadow detection methods can be exploited in order to obtain additional geometric and semantic cues about shape and position of its casting object ('shape from shadows') as well as the localization of the light source. While in the second one, the main purpose is usually change detection, scene matching or surveillance (usually in a background subtraction context). Shadows can in fact modify in a negative way the shape and color of the target object and therefore affect the performance of scene analysis and interpretation in many applications. This chapter wills mainly reviews shadow detection methods as well as their

A. Amato (✉) · M. G. Mozerov · F. X. Roca · J. Gonzàlez
Computer Vision Center, Universitat Autònoma de Barcelona, Cerdanyola del Vallès, Spain
e-mail: aamato@cvc.uab.es

M. G. Mozerov
e-mail: mozerov@cvc.uab.es

F. X. Roca
e-mail: xavir@cvc.uab.es

J. Gonzàlez
e-mail: poal@cvc.uab.es

I. Huerta
Institut de Robòtica i Informàtica industrial, Universitat Politècnica de Catalunya, Barcelona, Spain
e-mail: ihuerta@iri.upc.edu

Augment Vis Real (2014) 6: 23–47

DOI: 10.1007/8612_2012_3

Published Online: 12 September 2012

taxonomies related with the second case, thus aiming at those shadows which are associated with moving objects (moving shadows).

1 Introduction

Video Surveillance has been in our society for a long time [6, 30]. It began in the twentieth century to assist prison officials in the discovery of escape methods. However, it was not until the late-twentieth century that surveillance expanded to include the security of property and people. Video surveillance is more prevalent in Europe than anywhere in the world. For instance, in the past decade, successive UK governments have installed over 2.4 million surveillance cameras (about one for every 14 people).[1] The average Londoners are estimated to have their picture recorded more than three hundred times a day.[2] Traditionally video surveillance was used to display images on monitors inspected by guards or operators. This fact has allowed the observation of an increase number of places using less people and also to perform patrolling duties from the safety of a control room. However, a single operator can only monitor a limited amount of scenes simultaneously and for a limited amount of time, because the process of manual surveillance is very time-consuming and is a really tedious task.

The new breakthroughs in technology have led to a new generation of video surveillance. The current generation of video surveillance systems uses digital computing and communication technologies to improve the design of the original architecture, with the ultimate goal to create an automatic video surveillance system.

Recent trends in computer vision has delved into the study of cognitive vision systems, which uses visual information to facilitate a series of tasks on sensing, understanding, reaction and communication. In other words, video surveillance systems aim to automatically identify people, objects or events of interest in different kinds of environments. Although video surveillance is probably one of the most popular areas for research in the field of computer vision, and much effort has been made to achieve an automatic system, this goal has yet to be reached.

Nowadays, the task of a video surveillance system aims to provide support to the human operator. The system warns an operator when an event, e.g., possible risks or potential dangerous situations, is detected. Despite the fact that the long-term goal is to build a completely automated systems, the short-term one is to increase the robustness of the current systems in order to reduce false alarms. This goal can only be achieved if the systems are able to interpret the interaction of events in the scene. To reach this goal low and high level tasks must be performed.

---

1 http://news.bbc.co.uk/2/hi/uk_news/6108496.stm

2 http://epic.org/privacy/surveillance/

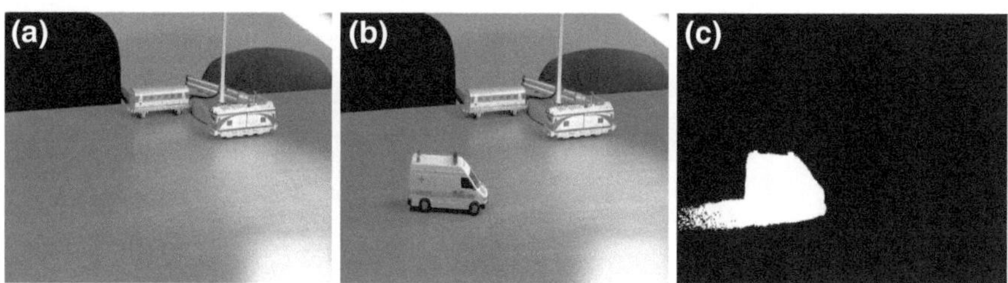

Fig. 1 Motion segmentation: **a** background image; **b** current image; and **c** segmented image

A method that is able to perform the low level tasks, namely detection, localization and tracking with high accuracy, can highly benefit the process of scene understanding.

In video surveillance moving object detection plays an important role [14, 43]. Along decades, different methods have been developed to extract moving region in the scene. However, the most common, simple and effective approach to moving object segmentation is Background Subtraction, where a stationary camera is used to observe dynamic events in a scene.

In moving object detection algorithms, moving cast shadows have a high probability to be misclassified as moving objects (foregrounds). Such an error is due to the fact that a moving object and its moving shadow share similar motion characteristics. An example of motion segmentation image based on background subtraction process is shown in Fig. 1c. The segmented image shows that the shadow was also segmented as a part of the object (foreground).

A shadow is a photometric phenomenon that occurs when an object partially or totally blocks the direct light source. Shadows can take any size and shape. In general, shadows can be divided into two major classes: self and cast shadows. A self shadow occurs in the portion of an object that is not illuminated by direct light. Cast shadows are the areas projected on a surface in the direction of direct light. Cast shadows can be further classified into umbra and penumbra. The region where the direct light source is totally blocked is called the umbra, while the region where it is partially blocked is known as the penumbra. These definitions are visually represented in Fig. 2.

Shadows in images are generally divided into static and dynamic shadows (see Fig. 3). Static shadows are shadows due to statics objects such as building, parked cars, trees, etc. Moving object detection methods do not suffer from static shadows since static shadows are modeled as a part of background. In contrary, dynamic (moving) shadows, the subject of interest in this chapter, are harmful for moving object detection methods. Shadows can be either in contact with the moving object, or disconnected from it (see Fig. 4). In the first case, shadows distort the object shape, making the use of subsequent shape recognition methods less reliable. In the second case, the shadows may be wrongly classified as an object in the scene. Typical problems caused by moving shadows in surveillance scenarios are shown in Fig. 5. In Fig. 5I, a traffic surveillance scene, shadows cause merging of

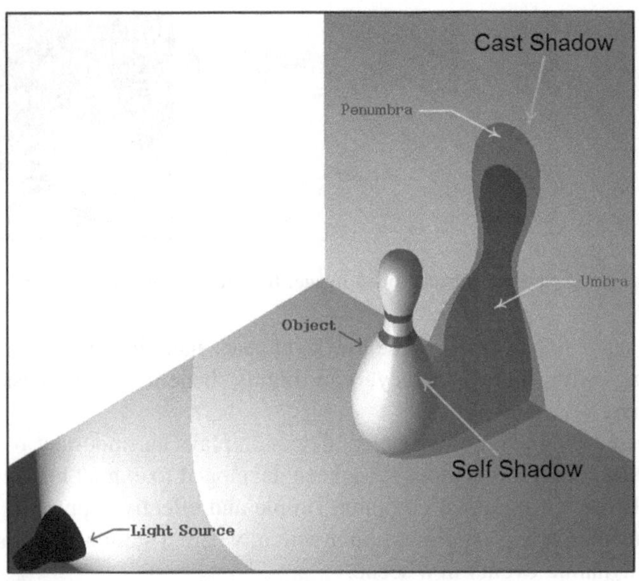

Fig. 2 Shadows types: self and cast (*Umbra* and *Penumbra*)

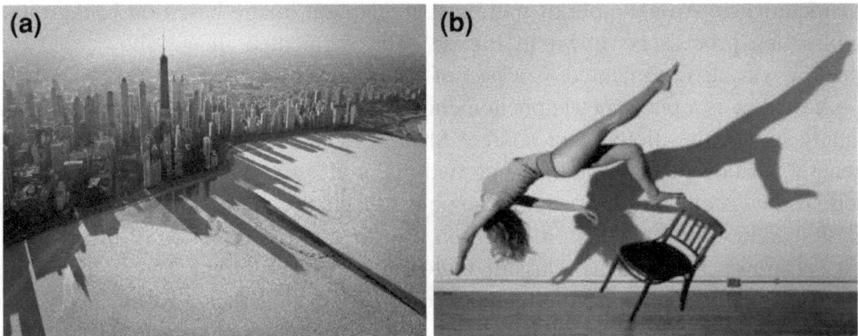

Fig. 3 Shadows from static and moving objects. **a** Shadows casted by static objects. **b** Shadow casted by a moving object

multiple objects; in Fig. 5II, an indoor scenario, shadows are projected on the floor and on the wall. In this case a false positive foreground (shadow casted on the wall) occurs; and in Fig. 5III, a long shadow causes a severe object shape distortion in an outdoor scenario. Clearly, in many image analysis applications, the existence of moving cast shadows may lead to an inaccurate object segmentation. Consequently, tasks such as object description and tracking are severely affected, thus inducing an erroneous scene analysis.

This chapter is organized in the following Sections. Section 2 some relevant background subtraction techniques are revised. Reported taxonomies on moving

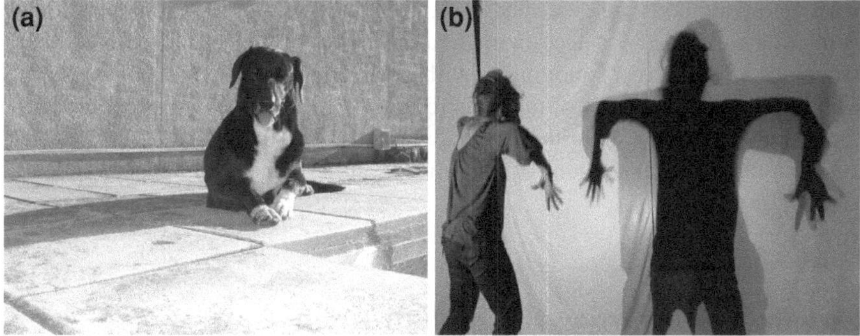

Fig. 4 Shadows location: **a** shadow is spatially connected to the object; **b** shadow is spatially unconnected to the object

(I)

(II)

(III)

Fig. 5 Negative effect of shadow in surveillance scenarios

cast shadow methods are described in Sect. 3. Section 4 gives a literature review on moving cast shadow detection methods. A discussion of open issues and main difficulties in the area of moving cast shadow detection is presented in Sect. 5. Section 6 gives an analysis of the tools for the performance evaluation on moving cast shadow detection algorithms. Finally, Sect. 7 briefly reviews the topics discussed in the different sections of this chapter and establishes the final concluding remarks of this work.

2 Methods for Moving Region Extraction

Detecting regions that correspond to moving objects such as vehicles and people in natural scenes is a significant and difficult problem for many vision-based applications. The extraction of the moving region is the first step to locate where a moving shadow can be detected.

The most used techniques for motion segmentation are: (i) background subtraction, (ii) frame differencing, (a combination of both), or (iii) optical flow. Even though many algorithms have been proposed in the literature [16, 41, 65], the problem of identifying moving objects in complex environment is still far from being completely solved.

Motion segmentation based on optical flow [5, 40] uses characteristics of flow vectors of moving objects over time to detect change regions in an image sequence. These methods can segment moving objects in video sequences even from a moving camera. However, most of these methods are computationally highly expensive and very sensitive to noise.

Temporal differencing technique attempts to extract moving regions by making use of a pixel-by-pixel difference between consecutive frames in a video sequence [55, 57]. It is very adaptive to dynamic scene changes. Nevertheless, it generally fails to extract the entire relevant pixels of moving objects.

Background subtraction is the most commonly used technique for motion segmentation in static scenes [4, 27, 38, 45, 53].

Basically, the methodology behind any background subtraction technique consists in subtracting a model of the static scene 'background' from each frame of a video sequence. (see Fig. 6). In general, a background subtraction technique can be divided into three phases: first, the generation of a suitable reference model, normally called background (training phase); second, the measurement procedure or classification (running phase) and finally; the model maintenance (updating phase). For each of these phases, particular challenging exist [24].

There are a large number of different algorithms using this background subtraction scheme. Nonetheless, they differ in: (i) the type of cues or structures employed to build the background representation; (ii) the procedure used for detecting the foreground region; and (iii) the updating criteria of the background model.

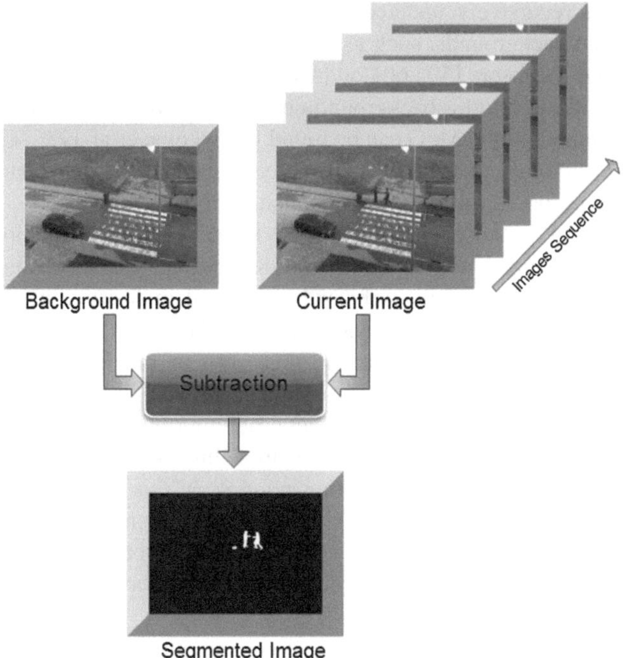

Fig. 6 Background subtraction representation

A naive version of the background subtraction scheme is employed by Heikkila and Silven [19], which classifies an input pixel as foreground if its value is over a predefined threshold when subtracted from the background model. This approach updates the background model in order to guarantee reliable motion detection using a first order recursive filter. However, this method is extremely sensitive to changes of dynamic scenes such as gradual illumination variation or physical changes such as *ghosts* (i.e., when an object already represented in the background model begins to move). In order to overcome these difficulties, statistical approaches have been applied [64]. These approaches make use of statistical properties of each pixel (or regions), which are updated dynamically during all the process in order to construct the background model.

Haritaoglu et al. in W^4 [18] apply background subtraction by computing for each pixel in the background model, during a training period, three values: its minimum and maximum intensity values, and the maximum intensity difference between consecutive frames. Background model pixels are updated using pixel-based and object-based updating conditions to be adaptive to illumination and physical changes in the scene. However, this approach is rather sensitive to shadows and lighting changes, since the only cue used is intensity.

Alternatively, Wren et al. in Pfinder [67] proposed a framework in which each pixel's value (in YUV space) is represented with a single Gaussian. Then, model parameters are recursively updated. However, a single Gaussian model cannot handle multiple backgrounds, such as waving trees.

Stauffer and Grimson [59, 60] addressed this issue by using a Mixture of Gaussians (MoG) to build a background color model for every pixel.

An improvement of the MoG can be found in Zivkovic et al. [73, 74], where the parameters of a MoG model are constantly updated, while selecting simultaneously the appropriate number of components for each pixel.

Elgammal et al. [13] use a non-parametric Kernel Density Estimation (KDE) to model the background. Their representation samples an intensity values for each pixel to estimate the probability of newly observed intensity values. The background model is also continuously updated to be adaptive to background changes. In addition to color-based information, their system incorporates region-based scene knowledge for matching nearby pixel locations. This approach can successfully handle the problem of small background motion such as tree branches.

Mittal et al. [40] use adaptive KDE for modeling background in motion, and implement optical flow to detect moving regions. In this way, their approach is able to manage complex background; however, the computational cost of this approach is quite high. Chen et al. [9] combine pixel- and block-based approaches to model complex background. Nevertheless, the method is very sensitive to camouflages and shadows.

Cheng et al. in [10] propose an on-line learning method which is able to work in real-time and can be implemented in GPU, which also gives similar results managing complex background. In [3] Barnich and Droogenbroeck also present a really fast method that can cope with background in motion and bootstrapping problems. The method adopts the idea of sampling the spatial neighborhood for refining the per-pixel estimation. The model updating relies on a random process that substitutes old pixel values with new ones. However, it cannot cope with camouflages and shadows. Another solution to bootstrapping problem is presented by Colombari et al. in [11], where a patch-based technique exploits both spatial and temporal consistency of the static background.

Li et al. [31] and Sheikh et al. [54] use Bayesian networks to cope with dynamic backgrounds. Li et al. uses a Bayesian framework that incorporates spectral, spatial, and temporal features to characterize background appearance. Sheik et al. apply non-parametric density estimation to model the background as a single distribution, thus handling multi-modal spatial uncertainties. Furthermore, they also use temporal information.

The use of layers for image decomposition based on the neighboring pixels is presented in [44]. They state that such approach is robust and efficient to handle dynamic backgrounds. Maddalena et al. [34] use neural networks to overcome the same problem. An improvement of it, using self organizing maps, can be found by Lopez-Rubio et al. [33], which can adapt its color similarity measure to the characteristics of the input video.

Mahadevan et al. in [35] uses a combination of the discriminant center-surround saliency framework with the modeling power of dynamic textures to solve problems with highly dynamic backgrounds and a moving camera. However, this method is not designed for high accurate segmentation.

Toyama et al. [62] in Wallflower use a three-component system to handle many canonical anomalies for background updating. Their work processes input images at various spatial scales, namely pixel, region, and frame levels. Reasonably good foreground detection can be achieved when moving objects or strong illumination changes (for example when turning on/off the light in an indoor scene) are present. However, it fails when modeling small motion in the background or local illumination variations.

Edge cues are also used for motion segmentation. Weiss [66] also extract intrinsic images using edge cues instead of color to obtain the reflectance image. This process requires several frames to determine the reflectance edges of the scene. A reflectance edge is an edge that persists throughout the sequence. Given reflectance edges, the approach re-integrates the information to derive a reflectance image. However, the reflectance image also contains scene illuminations because this approach requires prominent changes in the scene, specifically for the position of shadows.

Jabri et al. [25] use a statistical background modeling which combines color (in RGB space) with edges. The background model is computed in two distinct parts: the color model and the edge model. On the one hand, a color model is represented by two images, the mean and the standard deviation images. On the other hand, an edge model is built by applying the Sobel edge operator to each color channel, thereby yielding horizontal and vertical difference images. Subsequently, background subtraction is performed by subtracting the color and edge channels separately using confidence maps, and then combining the results to get the foreground pixels.

Javed et al. [26] present a method that uses multiple cues, based on color and gradient information. The approach tries to handle different difficulties, such as bootstrapping (initialization with moving objects), repositioning of static background objects, ghost and quick illumination changes using three distinct levels: pixel, region and frame level, inspired from [62].

At the pixel level, two statistical models of gradients and color based on mixture of Gaussians are separately used to classify each pixel as background or foreground. At the region level, foreground pixels obtained from the color model are grouped into regions, and the gradient model is then used to eliminate regions corresponding to highlights or ghosts.

Pixel-based models are updated based on decisions made at the region level. Lastly, the frame level ignores the color based subtraction results if more than 50 percent of the results are considered foreground, thereby using only gradient subtraction results to handle global illumination changes. Nevertheless, *ghosts* cannot be eliminated if the background contains a high number of edges.

Some of the aforementioned motion detection approaches generally obtain good segmentation in indoor and outdoor scenarios, thus some of them have been used in real-time surveillance applications for years. However, their performance is highly affected by the moving shadows.

3 Taxonomies of Moving Cast Shadow Detection Methods

Moving cast shadow detection algorithms are mainly based on the use of shadow descriptors. They basically model shadows by using properties such as: chromaticity invariant, textural patterns, photometric physical models, or even by analyzing the projected areas in term of size, shape and direction.

The methodology of moving cast shadow detection can further includes geometrical-shadow-information or spatial-shadow-cues as well as a training-shadow-stage, or a sort of combination of them. In turn, the methods can perform at different levels, considering only the information of a single pixel, using a set of pixels, or even performing with the information of the whole frame.

Diverse information that characterizes moving shadows is exploited and in many cases such information is combined or used in a different way. This fact makes very difficult to classify in a unique manner the moving cast shadow methods.

The main reported taxonomies in the literature were proposed by:

Prati et al. [47], present two layers taxonomy (*algorithm-based taxonomy*). The first layer classification considers whether the decision process introduces and exploits uncertainty. Deterministic approaches use an on/off decision process, whereas statistical approaches use probabilistic functions to describe the class membership. In turn, both layers are further divided. For statistical approaches the authors include parametric and non-parametric separation. In the case of deterministic methods, algorithms are classified by whether the decision is supported by model-based knowledge or not. Additionally, spectral, spatial and temporal information are also considered.

Salvador et al. [50] propose to divide shadow detection methods in (i) model-based methods and (ii) property-based methods. Model-based approaches work with models that represent a priori knowledge of the geometry of the scene, the object and the illumination. While property-based methods identify shadows by using properties such as the geometry, brightness and/or color of shadows.

Zhang et al. [72] describe moving cast shadow detection methods into: (i) color/spectrum-based methods; (ii) texture-based methods, and (iii) geometry-based methods. The color/spectrum-based methods attempt to describe the color change of a shadowed pixel and find the color feature that is illumination invariant. Texture-based methods consider that the texture of foreground object is different to the texture of background; while the texture of a shadowed area must be the same to the texture of background. Finally the geometric-based methods are focused on the characteristic of the casted shadow area. Usually the characteristics to be analyzed are direction, size and shape of the shadow. Often these methods can hardly be able to avoid the use of some prior knowledge of the scene. In turn, the authors also describe methods that make use of statistical inference of shadow models.

Ullab et al. [63] state that moving shadow removal methods can be partitioned into three categories: (i) intensity information, (ii) photometric invariant information and (iii) color and statistical information. The first classification concentrates in the brightness of the shadowed pixels. Typically a shadowed pixel decreases its brightness compared to the same pixel without shadow. The second classification includes those algorithms that exploit photometric-invariant-shadow property. Normally such photometric invariability can be obtained in normalized color spaces that can separately operate with the brightens and the chroma of the pixels. The last classification stands for methods which usually classify shadow by using statistical model of the pixel's information.

Sanin et al. [52] separate moving cast shadow removal methods into: (i) chromaticity-based methods; (ii) physical methods; (iii) geometrical-based methods and (iv) texture-based methods. Additionally, a secondary classification within each category is proposed.Chromacity-based methods are divided according their color space, level of granularity and additional spatial or temporal verification. Physical methods are divided according to their physical shadow model, learning algorithm and additional spatial or temporal cues. Geometry-based methods are divided according to their supported object type, whether they support multiple objects per blob, their main geometrical cue and additional cues. Texture-based methods are divided according to their weak shadow detector, texture correlation method and the size of the regions used in the correlation.

4 Methods for Moving Cast Shadow Detection

In this section the most classical and well known moving cast shadow detection approaches are presented.

Although, different taxonomies were revised in the previous section, we have organized this methods' review in pixel-level (those methods that perform with the information of a single pixel), region level (those methods that make use of a set of pixels) and frame-level[3] in order to unify as much as possible the different methods' characteristics.

Pixel-Level

Many shadow detection methods assume that a shadowed pixel becomes darker but with a similar chromaticity that the same pixel without shadow. Chromaticity is a measurement of color that is independent of intensity component. The invariability in chroma, between a (non-shadowed) pixel belonging to the background and the same (shadowed) pixel belonging to the current image, together with a brightness decrement, represent a distinctive shadow feature. Often methods that are using this shadow descriptor perform in color spaces where the distinction between brightness and chroma is supported. These common spaces are: HSV, HSI, YUV, C1C2C3, normalized RGB, etc (see Fig. 7).

[3] Frame-level methods are included, despite the fact that they are not widely used, in order to obtain a thorough review of the methods.

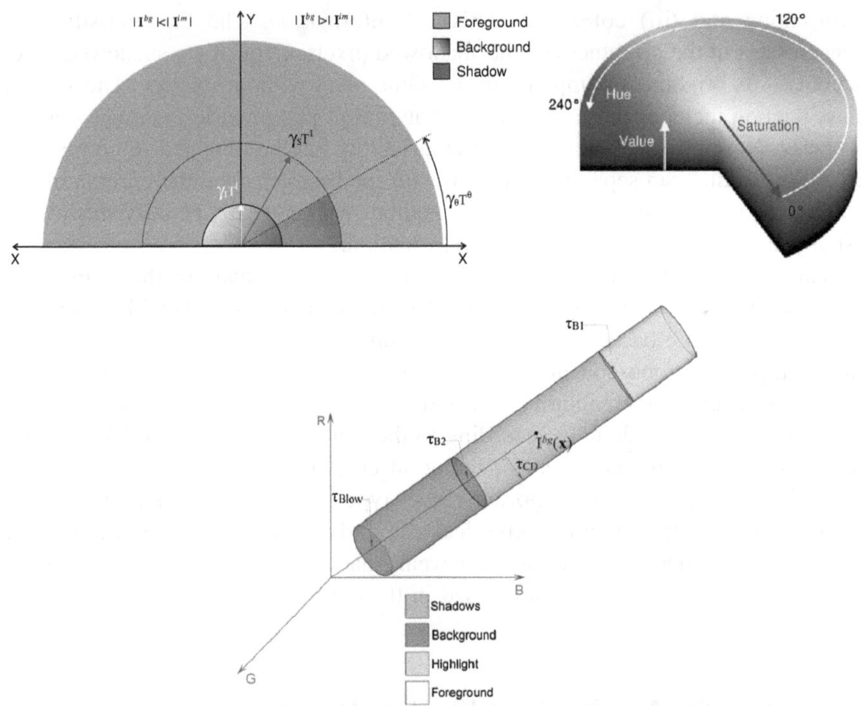

Fig. 7 Chromaticity invariant shadow descriptors in different color spaces. **a** Amato et al. [1], **b** Cucchiara et al. [12] and **c** Horprasert et al. [20]

For example, Cucchiara et al. [12] use shadow properties in the HSV color space to distinguish shadows from moving objects. These properties show that cast shadows darken the background in the luminance component, while the hue and saturation components change within certain limits.

Horprasert et al. [20] propose a color model that compares intensity to the chromaticity component at each pixel. Each pixel is classified as background, shaded, highlighted or moving foreground through a combination of three threshold values, which are defined over a single Gaussian distribution. An extension of this work based on multiple background pixels organized in a codebook is done by Kim et al. [28].

McKenna et al. [39] assume that cast shadows result in significant change in intensity without much change in chromaticity. Pixel's chromaticity is modeled using its mean and variance. In turn, the first-order gradient of each background pixel is also exploited. Moving shadows are then classified as background if the chromaticity or gradient information supports their classification as such.

The advantage of all the mentioned above methods reside in that they are fast (suitable for real-time applications), and easy to implement. However, they are specially restricted to achromatic shadows. Moreover, some of them often require explicit tuning of parameters for each scene.

There are some methods that aim to obtain an inference of the pixel values in the shadowed areas by using some photometric physical model. For such a purpose, a formulation can be achieved by exploiting: a reflectance model, an illumination model or an adaptation of classical color models. In order to obtain the appearance of the shadowed pixels some methods may need a training phase (it could be supervised or unsupervised), or/and some prior knowledge of the scene of interest.

Photometric physical model methods that implement statistical learning-based methodology have been developed to learn and remove cast shadows [22, 36, 37, 46]. For example, in the work of [37] a nonparametric framework to model surface behavior when shadows are cast on them is introduced. Physical properties of light sources and surfaces are employed in order to identify a direction in RGB space at which background surface values under cast shadows are found. However, these approaches are particularly affected by the training phase. These methods require a long training period.

In the work of Siala et al. [56] a statistical non-parametric shadow detection method is presented. First, in the learning phase, an image containing foreground, background and moving shadow is selected. The moving shadow regions are manually annotated. The information obtained from this annotation is then used to create a diagonal model that describes the shadow appearance in the RGB ratio color space. The shadow detection is obtained by performing a one class classification based on a support vector domain description (SVDD).

Region-Level

Although most of the methods that perform with a set of pixels typically make use of texture information, there are few methods that exploit other shadow descriptor such as: chromaticity invariant or photometric physical models.

Methods that use texture as shadow descriptor basically are based on the idea that a shadow is a semi-transparent region in the image. Thus, they assume that a strong correlation between two regions, one affected by shadow and the same region without the shadows effect, must exist. These methods try to obtain such a correlation using for example: local binary patterns (LBP), normalized cross-correlation (NCC), color cross covariant (CCC), Markov random field, etc. (see Fig. 8). For instance, Grest et al. [17] propose to tackle the detection of moving cast shadows using two similarity measurements, one is based on the normalized cross correlation (NCC) and the other is the color cross covariant (CCC). Basically the authors are interested in comparing pixel values at the same position in two images, (the current image and a reference image) and then inferring if there is a correlation between the information of these pixels. The computation of these measurements are done over a given window size. The NCC is calculated using the brightness of the pixel, while the CCC is obtained in the biconic Hue, Saturation and Lightness (HSL) color space. The authors assume that: (i) a shadowed pixel is darker than the corresponding pixel in the background image; (ii) the texture of the shadowed region is correlated with the corresponding texture of the background image. Despite the fact that CCC is used to solve the limitation of the method to

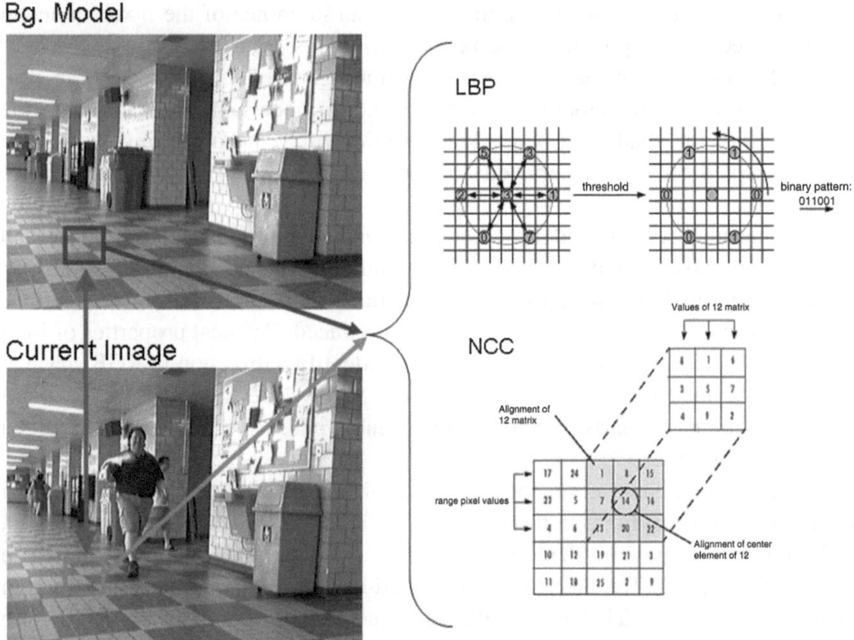

Fig. 8 Texture-oriented methodology

distinguish shadow from object over homogeneous areas, still the success of the approach under shadows-camouflage areas is far to be achieved.

Other approach based on NCC is proposed by Yuan et al. [71]. The authors proposed to include a multi-frame differencing strategy to improve the segmentation in those cases where the shadows cannot successfully be removed. This strategy is based on that shadowed regions differ a little in two consecutive frames. Therefore the biggest part of the shadows can be eliminated by frame difference, but only remain some shadow edges. These shadow edges are removed by using a new frame differencing step.

Jacques et al. [7] propose to detect shadows regions by using intensity measurement of a set of pixels. This measurement are computed by ratio-pixels (image/background) in a fixed 3×3 windows and the decision is based on a statistical non-parametric inference.

In the work of Yao et al. [70], textures are computed using the LBP combined with a RGB color model. The authors state that LBP can work robustly to detect moving shadows on rich texture regions. However, it fails when both the background image and the foreground objects share the same texture information. Therefore, to handle these situations, in this work the authors make use of a shadow invariant color distance in the RGB color space. They claims that pixel values changed due to shadows are mostly distributed along in the axis going toward the RGB origin point. Thus, they propose to compare the color difference between an observed color pixel and a background color pixel using their relative angle in RGB color space with respect to the origin and the changing range of the background color pixel up to last time instant.

Leone et al. [29] use a textural shadow descriptor by projecting the neighborhood of pixels onto a set of Gabor functions, extracted by applying a generalized scheme of the Matching Pursuit strategy. The methodology for shadow detection is based on the observation that shadows are half-transparent regions which retain the representation of the underlying background surface pattern. This approach assumes that shadow-regions contain same textural information, both in the current and in the background images.

In the work of Amato et al. [1] a method that introduces two discriminative features to detect moving cast shadow is presented. These features are computed based on angular and modular patterns, which are formed by similarity measurement between two sets of RGB color vectors. Unlike the most texture-based methods that often exploit spatial information, the patterns used in this approach are only photometric. This method could also be categorized as chromaticity invariant since it make uses of chroma and intensity information of a set of pixels to form a textural pattern.

Salvador et al. [49, 50] introduce a two stage method for segmenting moving shadows. The first stage segments the moving shadows in each frame of the sequence. In this stage the property that shadows casted on a surface reduce the surface intensities is exploited by using the photometric invariant C1C2C3 color space. In addition, to obtain a more robust result, the authors propose two schemes: (i) analyze a set of pixels (neighborhood) instead of a single pixel, and (ii) include geometrical verification based on boundary analysis of the shadow-candidate regions and testing the position of shadows with respect to objects. The second stage is used to obtain a coherent description of the segmented shadows over time. Therefore, the authors introduce a tracking shadow algorithm. An extension of this work was presented in [51] where the algorithm can segment cast shadow for both still and moving images.

Yang et al. [69] propose a moving cast shadow detection algorithm that combines shading, color, texture, neighborhoods and temporal consistency in the scene.

In comparison with methods that perform at the pixel-level, the aforementioned methods normally exploit texture information or use information from a set of pixels, making the detection more robust against noise and more efficient in those cases where ambiguity in the pixel's information occurs. However, the main drawback of these methods reside in the choice of the region's size that will be used. In other words, a strong dependency between the size of the region and the success of the method exists.

Many factors are involved in the choice of the region's size, for example: size of the object, textural composition of the background as well as of the object, etc. Consequently, an optimal region's size highly depends on the scene; moreover, the optimal size can change for different frames, even the optimal region size can changes within the frame. Furthermore, in many cases the computational time will vary with the size of such a region.

On the other hands, there are other region-based methods that perform with local adaptive regions. Basically they attempt to segment the moving area and then analyze and classify each segment based on shadow properties. These methods take advantage from pixel-level methods since they can make use of the information of a set of pixels. Additionally, they have also an advantage with respect to the fixed region-level methods since they can automatically adapt the area of analysis. A summary of this kind of methods is given below.

Toth et al. [61] propose a shadow detection algorithm based on color and shading information. They segment an image into several regions based on color information and the mean shift algorithm. They consider that the intensity values of a shadow pixel divided by the same pixel in the background image should be constant over a small segment.

In [15] an algorithm for outdoor scenarios is presented. Luminance, chrominance and gradient density information are exploited to create a shadow confidence score. Such a shadow score is based on three rules. The first rule claims that the luminance of the cast shadow is lower than the background. The second rule claims that the chrominance of the cast shadow is identical or slightly shifted when compared with background. And the last rule claims that the difference in gradient density between the cast shadow and background is lower than the difference in the distance of gradient between the object and background. The final classification combines the shadow score with a geometrical supporter. The geometrical cue used is based on the fact that the cast shadow is at the boundary region of moving foreground mask. That is, the cast shadow can be formed in any direction of the object, but not inside the object. However, the method is restricted to: (i) the areas where the shadows are casted on are not textured and (ii) the object shape is a convex hull which makes inappropriate to detect non-rigid object.

Rosin et al. [48] present a method based on the notion of a shadow as a semi-transparent region in the image which retains a (reduced contrast) representation of the underlying surface pattern, texture or gray value. The method uses a region growing algorithm which apply a growing criterion based on a fixed attenuation of the photometric gain over the shadow region, in comparison to the reference image. The problem with this approach is that region growing algorithm cannot perform accurately in the penumbra part of the shadow due to the intensity's variations inside of the shadow region.

Xu et al. [68] detect shadow region in indoor environment. The proposed method assumes that the shadow often appears around the foreground object. A number of techniques are used including initial change detection masks, Canny edge maps, multi-frame integration, edge matching, and conditional dilation. The method tries to detect shadow regions by extracting moving edges.

Chang et al. [8] propose a parametric Gaussian shadow model to detect and suppress pedestrian's shadow. The model makes uses of several features including the orientation, mean intensity, and center position being estimated from the properties of object movements.

In the work of Hsieh et al. [21] a line-based shadow modeling process is proposed to detect moving shadows in traffic surveillance. When a vehicle moves along a lane, it will have several boundary lines parallel or vertical to this lane. Then, the lane can provide useful information for shadow elimination and do not destroy vehicle shapes. In the method first all lanes dividing lines are detected. These lanes dividing lines from video sequences are detected by vehicle's histogram. This histogram is obtained by accumulating different vehicles' positions in a training period. According to these lines and their directions, two kinds of lines are used to eliminate shadows. The first one is the lines that are parallel to the dividing lines and the second one is the lines vertical to the dividing lines.

In [42] the authors propose an outdoor shadows removal method. It is based on a spatio-temporal-reflection test and a dichromatic reflection model. The approach is divided in several sequential steps. The step one starts with the motion mask, which is computed based on mixture of Gaussians. The intensity test takes the second step. This is in charge to discard all the foreground pixels that are more brightness than their corresponding background pixels. The third step so-called blue ratio test exploits the observation that shadows pixels falling on neutral surfaces, tend to be more bluish (this step can only be performed in neutral surfaces the authors propose to define a neutral surface based on the saturation level). The fourth step so-called albedo ratio segmentation performs a segmentation based on a spatio-temporal albedo ratio. Basically, this step attempts to obtain segmented regions with uniform reflectance. Step five removes the effect of the sky illumination. The authors claim that the reflection due to sky illumination (ambient reflection) is considered as an additive component; therefore they subtract the foreground pixels from the background. The regions that belong to the foreground will result with a very different color vectors that it is the contrary of the pixels belonging to the shadow regions. The last step aim to classify those regions that could not be labeled in previous stages. This stage computes the dominant color of the unclassified-regions (body color estimation) and compare with the body colors of material surfaces pre-stored as a background model (using a supervised-learning phase).

Similar to [42] Huerta et al. [23] use a multi-stage approach, however they use multiple cues: color and gradient information, together with known shadow properties. In this way, regions corresponding to potential shadows are grouped by considering the "bluish effect" and an edge partitioning. Additionally, temporal similarities between textures and spatial similarities between chrominance angle and brightness distortions are analyses for all potential shadows regions. Furthermore, geometrical shadow position is taken to account to avoid a misclassification of moving shadows.

In the method of Amato et al. [2], first an initial change detection mask containing moving objects and moving shadows is obtained using a background subtraction technique. Then, objects masks are computed by using connected component analysis. Based on the shadow luminance model, the authors state that in the luminance ratio space, a low gradient constancy exists in all shadowed regions, as opposed to foreground regions which, in most cases, exhibit higher gradients. To exploit these foreground-shadow characteristics, the authors designed a novel gradient-based segmentation algorithm to partition each object area into a set of low gradient segments (objects sub-segments). Then, objects sub-segments are classified as shadow or foreground, following three criteria: (i) luminance difference criterion; (ii) segment size criterion; and (iii) extrinsic terminal point weight criterion.

The challenge in these methods is not only in being able to properly analyze the segments, but also in the segmentation process. Nevertheless, this adaptive methodology is a promising way to detect moving cast shadows since the analysis is done with the context of the shadowed area having all the shadow information.

Frame-Level

There are a very few moving cast shadow detection methods that perform at the frame-level. Normally, these methods are not used in a background subtraction context. Hence, some of the closest methods related to the research line proposed in this work are briefly describe.

Liu et al. [32] detect shadows using pixel-level information, region-level information, and global-level information. Pixel-level information is extracted using GMM in the HSV color space. Local-level information is used in two ways. First, if a pixel gets a sample that is likely to be a shadow, then not only the GMM of that pixel is updated but the GMM of neighbor pixels is also updated. Second, Markov random fields are employed to represent the dependencies among neighboring pixels. For global-level information, statistical feature is exploited for whole scene over several consecutive frames.

Stauder et al. [58] use a physics-based luminance model to describe illumination changes. They assume a plain textured background and a cast shadow is determined by combining the results of change detection, static edge detection, shading change detection and penumbra detection.

5 Open Issues and Difficulties to Overcome

Shadows are normally considered as a local illumination problem. Obviously, areas affected by cast shadow experience a change of illumination. Often this illumination change is considered only as a decrease in brightness, without significant variation in chromaticity. However, the assumption that pixel's chromaticity is invariant to cast shadows is not always correct. It is correct, in fact, only when the chromatic components of the light sources are similar between them and there is no color blending among objects. This type of shadow is often called an achromatic shadow, while those that are not achromatic are referred to as chromatic shadows [2]. Removing chromatic shadows is a particularly challenging task due to the fact that they are extremely difficult to distinguish from the foreground because they have not a clearly defined photometric pattern. The interplay between color and texture in the background and shadows is highly variable and difficult to characterize. Another non trivial problem occurs when there is no difference in

Table 1 Qualitative evaluation for different methods

Methods	Chromatic shadows	Shadow camoufage	Surface topology
Cucchiara et al. [12]	*High*	*High*	*Low*
Horprasert et al. [20]	*High*	*High*	*Low*
McKena et al. [39]	*High*	*High*	*High*
Kim et al. [28]	*High*	*High*	*Low*
Siala et al. [56]	*Low*	*High*	*Low*
Martel-Brisson et al. [37]	*High*	*High*	*Medium*
Huang et al. [22]	*High*	*High*	*Medium*
Fung et al. [15]	*High*	*High*	*High*
Huerta et al. [23]	*Low*	*High*	*High*
Toth et al. [61]	*Low*	*Medium*	*High*
Nadimi et al. [42]	*Low*	*High*	*Medium*
Amato et al. [1]	*High*	*Medium*	*High*
Yuan et al. [71]	*Low*	*Medium*	*High*
Grest et al. [17]	*High*	*Medium*	*High*
Yao et al. [70]	*High*	*Medium*	*High*
Leone et al. [29]	*Low*	*High*	*High*
Jacques et al. [7]	*Low*	*High*	*High*
Yang et al. [69]	*Low*	*High*	*Medium*
Amato et al. [2]	*Low*	*Low*	*Medium*

The table valuates the negative effect degree with: *Low, Medium and High*

chromaticity between foreground object and background (e.g. black car is moving in highway), hence inducing a strong similarity between shadow-foreground pixels. Such effect is called as shadow camouflage. Despite of the fact that many articles of moving cast shadow detection have been published during the las years, only few works in the literature address these two major problems: chromatic shadow identification, and shadow detection in camouflaged areas. Although, methods that aim to extract certain patterns (region-based) make the detection more robust against noise and more efficient in those cases where ambiguity in the pixel's information occurs; these methods may also suffer from the chromatic shadow effect. Furthermore, an intrinsic limitation of some of these methods resides in the textural composition of the background as well as of the object. The surfaces' topology in terms of texture or texture-less plays a significant role in the patters extraction task.

Table 1 presents a qualitative comparison among several moving cast shadow detection algorithms. It reports the negative impact that chromatic shadow and shadow camouflage might cause over the performance of the methods. The table valuates the negative effect degree with: *Low, Medium and High.* In turn, the table also shows the dependency of algorithms' performance respect to surface topology (namely texture or texture-less). The degree of this dependency is similarly classified.

6 Evaluation of Moving Cast Shadow Detection Methods

This section explains the essential tools to evaluate the performance of moving cast shadow detection approaches. Commonly, moving cast shadow detection methods are evaluated using the metrics and the sequences described below.

Metrics

The quantitative comparison normally is based on two standard metrics for evaluating the performance of cast shadow detection algorithm introduced by Prati et al. [47]: shadow detection rate (η) and shadow discrimination rate (ξ). These two metrics are as follow:

$$\eta = \frac{TP_S}{TP_S + FN_S}; \qquad \xi = \frac{\overline{TP_F}}{TP_F + FN_F}. \tag{1}$$

where TP and FN stand for true positive and false negative pixels detected respect to both shadows S and foreground F. $\overline{TP_F}$ is the number of true positive foreground pixels detected minus the number of points detected as shadows but belonging to the foreground.

The shadow detection rate η is related to the percentage of shadow pixels correctly classified, while the shadow discrimination rate ξ is concerned with foreground pixels correctly classified.

Sequences

The typical sequences normally used for evaluating moving cast shadow detection methods are[4]:

1. (Highway II, Campus, Laboratory and Intelligent Room)[5]
2. (Hallway, HWI, HWIII)[6]
3. (CVC-Outdoor, Football Match, Pets-2009 View 7)[7]

The characteristics of the sequences are summarized in Table 2. The characteristics' description is organized in terms of: (i) Frames, (ii) Scene, (iii) Object and (iv) Shadows. Where (i) includes number of frames, hand-labeled (ground truth) frames and image size. The second category (ii) specifies the type, background and noise of the scene. The third category (iii) reports the classes and the sizes of the foreground objects. Finally, the last category (iv) describes the casted shadows on the scene in terms of size, visibility (referred to the perception of the human eye), direction, camouflage and chromatic effect.

---

[4] Note that for a quantitative evaluation a ground truth is necessary, the sequences as well as their ground truth are publicly accessible in the listed links.

[5] http://cvrr.ucsd.edu/aton/shadow/

[6] http://vision.gel.ulaval.ca/~CastShadows/

[7] http://www.cvc.uab.es/~aamato/Shadows_Detection/; http://www.cvg.rdg.ac.uk/PETS2009/a.html

Table 2 Description of typical sequences normally used for evaluating moving cast shadow detection methods

		Sequences			
		HighwayII	Campus	Laboratory	Intelligent Room
Frames	Number	*500*	*1179*	*887*	*300*
	Hand-labeled	*5*	*6*	*7*	*113*
	Size	*320x240*	*352x288*	*320x240*	*320x240*
Scene	Type	*Outdoor*	*Outdoor*	*Indoor*	*Indoor*
	Background	*Textured-less*	*Variable*	*Variable*	*Variable*
	Noise	*High*	*High*	*High*	*High*
Object	Class	*Vehicles*	*Vehicle and People*	*People*	*People*
	Size	*Variable*	*Large*	*Large*	*Large*
Shadows	Size	*Small*	*Large*	*Variable*	*Large*
	Visibility	*High*	*Low*	*Low*	*Low*
	Direction	*Single horizontal*	*Single horizontal*	*Multiple*	*Multiple*
	Camouflage	*High*	*Low*	*Low*	*Low*
	Chromatic effect	*High*	*Low*	*Low*	*Low*

		Sequences		
		Hallway	HWI	HWIII
Frames	Number	*1800*	*440*	*2227*
	Hand-labeled	*13*	*8*	*7*
	Size	*320x240*	*320x240*	*320x240*
Scene	Type	*Indoor*	*Outdoor*	*Outdoor*
	Background	*Textured*	*Textured-less*	*Texture-less*
	Noise	*Medium*	*Medium*	*Medium*
Object	Class	*People*	*Vehicles*	*Vehicles*
	Size	*Variable*	*Large*	*Variable (small)*
Shadows	Size	*Variable*	*Large*	*Variable*
	Visibility	*Low*	*High*	*High*
	Direction	*Multiple*	*Single horizontal*	*Single horizontal*
	Camouflage	*Low*	*High*	*High*
	Chromatic effect	*Low*	*Low*	*High*

		Sequences		
		CVC Outdoor	Football Match	Pets 2009 V7
Frames	Number	*800*	*2699*	*795*
	Hand-labeled	*12*	*13*	*16*
	Size	*320x240*	*320x240*	*720x576*
Scene	Type	*Outdoor*	*Outdoor*	*Outdoor*
	Background	*Textured*	*Textured-less*	*Variable*
	Noise	*Low*	*Medium*	*Low*
Object	Class	*People*	*People*	*People*
	Size	*Large*	*Small*	*Variable*
Shadows	Size	*Large*	*Small*	*Variable*
	Visibility	*High*	*Low*	*Low*
	Direction	*Single horizontal*	*Multiple horizontal*	*Single horizontal*
	Camouflage	*Low*	*Low*	*Low*
	Chromatic effect	*Medium*	*Low*	*Low*

7 Conclusion

Firstly, the problematic of moving cast shadows in video surveillance applications has been introduced in this chapter. Additionally, classical and well-known background subtraction methods were also described. Later on, a comprehensive survey of the most significant moving cast shadow approaches as well as their taxonomies has been presented. We have observed that moving cast shadow detection methods that only exploit chromaticity invariant property are not intrinsically prepared to cope with 'chromatic shadows'. In turn, methods that perform at the 'pixel level' highly decrease their performance in those cases where 'shadow camouflage' and 'chromatic shadows' occur, since the information of a single pixel is not enough to discriminate between shadow and foreground due to the ambiguity in their pixels values. In comparison with methods that perform at the pixel-level, the region-based method make the detection more robust against noise and more efficient in those cases where ambiguity in the pixel's information occurs. However, these methods may also suffer from the chromatic shadow effect. Furthermore, an intrinsic difficulty of some of these methods (fixed region-based) resides in the criterion of the region's size that is used. Thus a strong dependency between the size of the region and the success of the method exists. Several factors are involved in the choice of the region's size, for example: size of the object, textural composition of the background as well as of the object, etc. Consequently, an optimal region's size is highly depending on the scene; moreover, an optimal size can change in different frames of the same scene or even the optimal region's size can change within the frame. Finally, the metric and the most employed surveillance data sets to evaluate the performance of moving cast shadow algorithms were reported.

Acknowledgments Consolider-Ingenio 2010: MIPRCV (CSD200700018); Avanza I+D ViCoMo (TSI-020400-2009-133) and DiCoMa (TSI-020400-2011-55); along with the Spanish projects TIN2009-14501-C02-01 and TIN2009-14501-C02-02.

References

1. Amato, A., Mozerov, M., Roca, X., Gonzàlez, J.: Robust real-time background subtraction based on local neighborhood patterns. In: EURASIP Journal on Advances in Signal Processing, pp. 1–7, June 2010
2. Amato, A., Mozerov, M.G., Bagdanov, A.D., Gonzàlez, J.: Accurate moving cast shadow suppression based on local color constancy detection. Image Process. IEEE Trans. **20**(10), 2954–2966 (2011)
3. Barnich, O., Van Droogenbroeck, M.: Vibe: a universal background subtraction algorithm for video sequences. IEEE TIP **20**(6), 1709–1724 (2011)
4. Brutzer, S., Hoferlin, B., Heidemann, G.: Evaluation of background subtraction techniques for video surveillance. In: IEEE CVPR'11, pp. 1937–1944, June 2011
5. Bugeau, A., Perez, P.: Detection and segmentation of moving objects in highly dynamic scenes. In: IEEE CVPR'07, pp. 1–6, June 2008

6. Caputo, A.: Digital Video Surveillance and Security. Butterworth-Heinemann, Burlington (2010)
7. Cezar Silveira Jacques, J., Rosito Jung, C., Musse, S.R.: A background subtraction model adapted to illumination changes. In: Image Processing, 2006 IEEE International Conference on, pp. 1817–1820, October 2006
8. Chang, C.-J., Hu, W.-F., Hsieh, J.-W., Chen, Y.-S.: Shadow elimination for effective moving object detection with gaussian models. In: Pattern Recognition, 2002. Proceedings. 16th International Conference on, vol. 2, pp. 540–543, 2002
9. Chen, Y., Chen, C., Huang, C., Hung, Y.: Efficient hierarchical method for background subtraction. Pattern Recognit. **40**(10), 2706–2715 (2007)
10. Cheng, L., Gong, M., Schuurmans, D., Caelli, T.: Real-time discriminative background subtraction. IEEE TIP **20**(5), 1401–1414 (2011)
11. Colombari, A., Fusiello, A., Murino, V.: Patch-based background initialization in heavily cluttered video. IEEE TIP **19**(4), 926–933 (2010)
12. Cucchiara, R., Grana, C., Piccardi, M., Prati, A., Sirotti, S.: Improving shadow suppression in moving object detection with HSV color information. In: Intelligent Transportation Systems, 2001. Proceedings. 2001 IEEE, pp. 334–339, 2001
13. Elgammal, A., Harwood, D., Davis, L.S.: Nonparametric background model for background subtraction. In: ECCV'00, pp. 751–767, Dublin, 2000
14. Forsyth, D., Ponce, J.: Computer Vision: A Modern Approach. Prentice Hall, Upper Saddle River (2002)
15. Fung, G.S.K., Yung, N.H.C., Pang, G.K.H., Lai, A.H.S.: Effective moving cast shadow detection for monocular color image sequences. In: Image Analysis and Processing, 2001. Proceedings. 11th International Conference on, pp. 404–409, 2001
16. Gavrila, D.M.: The visual analysis of human movement: a survey. Comput. Vis. Image Underst. **73**, 82–98 (1999)
17. Grest, D., Frahm, J.M., Koch, R.: A color similarity measure for robust shadow removal in real time. In: Vision, Modeling and Visualization, pp. 253–260, 2003
18. Haritaoglu, I., Harwood, D., Davis, L.S: W4: real-time surveillance of people and their activities. IEEE TPAMI **22**(8), 809–830 (2000)
19. Heikkila, J., Silven, O.: A real-time system for monitoring of cyclists and pedestrians. In: Proceedings of the Second IEEE Workshop on Visual Surveillance, pp. 74–81, Washington, DC, 1999. IEEE Computer Society, Washington (1999)
20. Horprasert, T., Harwood, D., Davis, L.S.: A statistical approach for real-time robust background subtraction and shadow detection. In: ICCV Frame-Rate WS. IEEE, 1999
21. Hsieh, J.-W., Yu, S.-H., Chen, Y.-S., Hu, W.-F.: A shadow elimination method for vehicle analysis. In: Pattern Recognition, 2004. ICPR 2004. Proceedings of the 17th International Conference on, vol. 4, pp. 372–375, 2004
22. Huang, J.-B., Chen, C.-S.: Moving cast shadow detection using physics-based features. In: Computer Vision and Pattern Recognition, IEEE Computer Society Conference on, pp. 2310–2317, 2009
23. Huerta, I., Holte, M., Moeslund, T.B., Gonzàlez, J.: Detection and removal of chromatic moving shadows in surveillance scenarios. In: ICCV2009, Kyoto, 2009
24. Huerta, I., Amato, A., Roca, F.X., Gonzàlez, J.: Multiple cues fusion for robust motion segmentation using background subtraction. Neurocomputing, Elsevier, (2011, in press)
25. Jabri, H.W.S., Duric, Z., Rosenfeld, A.: Detection and location of people in video images using adaptive fusion of color and edge information. In: 15th ICPR, vol. 4, pp. 627–630, Barcelona, Sept 2000
26. Javed, O., Shafique, K., Shah, M.: A hierarchical approach to robust background subtraction using color and gradient information. In: Proceedings of the Workshop on Motion and Video Computing (MOTION'02), p. 22, Orlando, 2002
27. Karaman, M., Goldmann, L., Yu, D., Sikora, T.: Comparison of static background segmentation methods. In: VCIP '05, Beijing, July 2005

28. Kim, K., Chalidabhongse, T.H., Harwood, D., Davis, L.S: Real-time foreground-background segmentation using codebook model. Real-Time Imaging **11**(3), 172–185 (2005)
29. Leone, A., Distante, C.: Shadow detection for moving objects based on texture analysis. Pattern Recognit. **40**(4), 1222–1233 (2007)
30. Lin, W. (ed.): Video Surveillance. InTech, (2011). ISBN 978-953-307-436-8
31. Li, L., Huang, W., Gu, I.Y.-H., Tian, Q.: Statistical modeling of complex backgrounds for foreground object detection. IEEE TIP **13**(11), 1459–1472 (2004)
32. Liu, Z., Huang, K., Tan, T., Wang, L.: Cast shadow removal combining local and global features. In: Computer Vision and Pattern Recognition, 2007. CVPR '07. IEEE Conference on, pp. 1–8, June 2007
33. Lopez-Rubio, E., Luque-Baena, R.M., Dominguez, E.: Foreground detection in video sequences with probabilistic self-organizing maps. Int. J. Neural Syst. **21**(3), 225–246 (2011)
34. Maddalena, L., Petrosino, A.: A self-organizing approach to background subtraction for visual surveillance applications. IEEE TIP **17**(7), 1168–1177 (2008)
35. Mahadevan, V., Vasconcelos, N.: Spatiotemporal saliency in dynamic scenes. IEEE TPAMI **32**(1), 171–177 (2010)
36. Martel-Brisson, N., Zaccarin, A.: Learning and removing cast shadows through a multidistribution approach. Pattern Anal. Mach. Intell. IEEE Trans. **29**(7), 1133–1146 (2007)
37. Martel-Brisson, N., Zaccarin, A.: Kernel-based learning of cast shadows from a physical model of light sources and surfaces for low-level segmentation. In: CVPR08, pp. 1–8, 2008
38. McIvor, A.: Background subtraction techniques. In: Proceedings of Image and Vision Computing, Auckland, 2000
39. McKenna, S.J., Jabri, S., Duric, Z., Rosenfeld, A., Wechsler, h.: Tracking groups of people. Comput. Vis. Image Underst. **80**(1), 42–56 (2000)
40. Mittal, A., Paragios, N.: Motion-based background subtraction using adaptive kernel density estimation. In: Proceedings of CVPR'04, vol. 2, pp. 302–309, Washington DC, July 2004
41. Moeslund, T.B., Granum, E.: A survey of computer vision-based human motion capture. Comput. Vis. Image Underst. **81**(3), 231–268 (2001)
42. Nadimi, S., Bhanu, B.: Physical models for moving shadow and object detection in video. Pattern Anal. Mach. Intell. IEEE Trans. **26**(8), 1079–1087 (2004)
43. Obinata, G., Dutta, A.: Vision Systems: Segmentation and Pattern Recognition. I-Tech Education and Publishing, Vienna (2007)
44. Patwardhan, K.A., Sapiro, G., Morellas, V.: Robust foreground detection in video using pixel layers. IEEE TPAMI **30**(4), 746–751 (2008)
45. Piccardi, M.: Background subtraction techniques: a review. In: IEEE International Conference on Systems, Man and Cybernetics, vol. 4, pp. 3099–3104, The Hague, 2004
46. Porikli, F., Thornton, J.: Shadow flow: a recursive method to learn moving cast shadows. In: Computer Vision, 2005. ICCV 2005. Tenth IEEE International Conference on, vol. 1, pp. 891–898, 2005
47. Prati, A., Mikic, I., Trivedi, M.M., Cucchiara, R.: Detecting moving shadows: algorithms and evaluation. IEEE Trans. Pattern Anal. Mach. Intell. **25**(7), 918–923 (2003)
48. Rosin, P., Ellis, T.: Image difference threshold strategies and shadow detection. In: Proceedings of British Machine Vision Conference, pp. 347–356. BMVA Press, Surrey (1995)
49. Salvador, E., Cavallaro, A., Ebrahimi, T.: Shadow identification and classification using invariant color models. In: Acoustics, Speech, and Signal Processing, 2001. Proceedings. (ICASSP '01). 2001 IEEE International Conference on, vol. 3, pp. 1545–1548, 2001
50. Salvador, E., Cavallaro, A., Ebrahimi, T.: Spatio-temporal shadow segmentation and tracking. In: Proceedings of Visual Communications and Image Processing, pp. 389–400, 2003
51. Salvador, E., Cavallaro, A., Ebrahimi, T.: Cast shadow segmentation using invariant color features. Comput. Vis. Image Underst. **95**(2), 238–259 (2004)
52. Sanin, A., Sanderson, C., Lovell, B.C.: Shadow detection: a survey and comparative evaluation of recent methods. Pattern Recogn. **45**(4), 1684–1695 (2012)

53. SanMiguel, J.C., Martinez, J.M.: On the evaluation of background subtraction algorithms without ground-truth. In: Advanced Video and Signal Based Surveillance (AVSS), 2010 Seventh IEEE International Conference on, pp. 180–187, Sept 2010
54. Sheikh, Y., Shah, M.: Bayesian modeling of dynamic scenes for object detection. IEEE TPAMI **27**(11), 1778–1792 (2005)
55. Shen, J.: Motion detection in color image sequence and shadow elimination. Vis. Commun. Image Process. **5308**, 731–740 (2004)
56. Siala, K., Chakchouk, M., Chaieb, F., Besbes, O.: Moving shadow detection with support vector domain description in the color ratios space. In: Pattern Recognition, 2004. ICPR 2004. Proceedings of the 17th International Conference on, vol. 4, pp. 384–387, August 2004
57. Spagnolo, P., Orazio, T.D., Leo, M., Distante, A.: Moving object segmentation by background subtraction and temporal analysis. Image Vis. Comput. **24**(5), 411–423 (2006)
58. Stauder, J., Mech, R., Ostermann, J.: Detection of moving cast shadows for object segmentation. IEEE Trans. Multimedia **1**(1), 65–76 (1999)
59. Stauffer, C., Grimson, W.E.L.: Adaptive background mixture models for real-time tracking. In: IEEE CVPR'99, vol. 1, pp. 22–29, Ft. Collins, 1999
60. Stauffer, C., Eric, W., Grimson, L.: Learning patterns of activity using real-time tracking. IEEE TPAMI **22**(8), 747–757 (2000)
61. Toth, D., Stuke, I., Wagner, A., Aach, T.: Detection of moving shadows using mean shift clustering and a significance test. In: International Conference on Pattern Recognition (ICPR 2004), vol. 4, pp. 260–263, 2004
62. Toyama, K., Krumm, J., Brumitt, B., Meyers, B.: Wallflower: principles and practice of background maintenance. In: Proceedings of ICCV'99, vol. 1, pp. 255–261, Kerkyra, 1999
63. Ullah, H., Ullah, M., Uzair, M., Rehman, F.: Comparative study: the evaluation of shadow detection methods. Int. J. Video Image Process. Netw. Secur. **10**(2), 1–7
64. Wang, L., Hu, W., Tan, T.: Recent developments in human motion analysis. Pattern Recognit. **36**(3), 585–601 (2003)
65. Wang, L., Hu, W., Tan, T.: Recent developments in human motion analysis. Pattern Recognit. **36**(3), 585–601 (2003)
66. Weiss, Y.: Deriving intrinsic images from image sequences. In: Proceedings of ICCV'01, vol. 02, pp. 68–75, Vancouver, 2001
67. Wren, C.R., Azarbayejani, A., Darrell, T., Pentland, A.P: Pfinder: real-time tracking of the human body. IEEE TPAMI **19**(7), 780–785 (1997)
68. Xu, D., Li, X., Liu, Z., Yuan, Y.: Cast shadow detection in video segmentation. Pattern Recognit. Lett. **26**(1), 91–99 (2005)
69. Yang, M.-T., Lo, K.-H., Chinag, C.-C., Tai, W.-K.: Moving cast shadow detection by exploiting multiple cues. Image Process. IET **2**(2), 95–104 (2008)
70. Yao, J., Odobez, J.M.: Multi-layer background subtraction based on color and texture. In: IEEE CVPR'07, pp. 17–22, Minneapolis, June 2007
71. Yuan, C., Yang, C., Xu, Z.: Simple vehicle detection with shadow removal at intersection. In: Proceedings of the 2010 Second International Conference on Multi-Media and Information Technology, volume 02 of MMIT '10, pp. 188–191. IEEE Computer Society, 2010
72. Zhang, W., Wu, Q.M.J., Fang, X.: In: Obinata, G., Dutta, A. (eds.): Vision Systems: Segmentation and Pattern Recognition. Moving Cast Shadow Detection. InTech, (2007)
73. Zivkovic, Z.: Improved adaptive gaussian mixture model for background subtraction. In: Proceedings of ICPR'04, vol. 2, pp. 23–26, August 2004
74. Zivkovic, Z., Heijden, F.: Efficient adaptive density estimation per image pixel for the task of background subtraction. Pattern Recognit. Lett. **27**(7), 773–780 (2006)

Moving Object Detection and Tracking in Wide Area Motion Imagery

Varun Santhaseelan and Vijayan K. Asari

Abstract Tracking objects in wide area imagery is a difficult task because of very low image resolution to even detect the presence of small objects in a scene. This chapter presents a new methodology for moving object detection and tracking in such low resolution wide area imagery that can work well in varying conditions as well as for very small objects like pedestrians. The basic concept that has been used for the development of this algorithm is that all the information that is available about the object of interest has to be utilized in the tracking process. As we are considering only the moving objects, there is no need to consider stationary structures like buildings, trees and other landmarks in the context of tracking. This has motivated for the development of a feature based tracking mechanism, which makes use of a dense version of localized histogram of oriented gradients on the difference images. A basic Kalman filter is used as the predictive mechanism in the tracking methodology. The robustness of this algorithm is illustrated by tracking various objects of interest in varying situations. The effect of image enhancement algorithms in relation with tracking in shadows is also illustrated with experimental results. It is observed that super-resolution algorithms can play an important role in improving tracking in long range videos.

V. Santhaseelan · V. K. Asari (✉)
Department of Electrical and Computer Engineering, University of Dayton, Dayton, OH 45469, USA
e-mail: vasari1@udayton.edu

V. Santhaseelan
e-mail: santhaseelanv1@udayton.edu

Augment Vis Real (2014) 6: 49–70
DOI: 10.1007/8612_2012_9

Published Online: 12 September 2012

1 Introduction

Even after decades of research, object tracking still remains an active problem in the field of computer vision. With the advancement in sensor technology, it has become possible to have very large geographical areas under surveillance. However, the tracking methodologies that have been perfected for short range videos cannot be applied on videos obtained from wide area surveillance cameras. This is because of the low resolution of the objects of interest coupled with low frame rate of video and large shifts in video frames because of the movement of the aerial vehicle that captures the video. The objects that can be clearly observed on the scene are buildings and trees while cars and other vehicles appear miniscule and are of the size of tens of pixels in the frame. The lack of sufficient usable data poses a challenge to current detection and tracking algorithms. A sample image of the data that has been used in our research and some random objects in the image are shown in Fig. 1.

The current state of object tracking algorithms is explained in detail in [1]. There is a multitude of perspectives from which the problem of object tracking can be viewed with each perspective having its own advantages and disadvantages. However, the existing state of the art techniques fail miserably when it comes to long range tracking where the object of interest has a total size of ten pixels. Point features [2–5] can be used to track objects over subsequent frames. However, in the case of objects in wide area imagery, the number of point features that can be extracted is very less as illustrated in Fig. 2.

One of the methods to have a robust tracking mechanism using features is to have color features and textural details being used as features to track objects as presented in [6]. However, we have used the data from a monochromatic sensor and not much textural details are available on the object to be tracked. The lack of detail also affects the performance of appearance model based tracking of pedestrians [7].

This chapter tackles the problem of long range tracking using a dense feature set used in conjunction with the Kalman tracker. The main idea is to use all available information to describe a feature set that can best describe the object of interest. The fact that the object of interest will be in a state of motion for most of the time is an added advantage when it comes to tracking very small objects. The main contributions of this chapter are: (a) development of an efficient framework for tracking objects of various sizes in wide area motion imagery, and (b) development of a framework to track objects with the aid of image enhancement and super-resolution in wide area motion imagery.

For the experiments illustrated in this chapter, we have used the publicly available Columbus Large Image Format (CLIF) database [8]. This database is a collection of images of downtown Columbus which have been collected from an aerial vehicle at an altitude of almost 7,000 feet. The data was collected using six cameras oriented at different angles to have maximum coverage of the area while making sure that there is sufficient overlap between images to aid in the process of registering those images. For this chapter, we have not considered stitching the data from different cameras together. Our focus is on tracking various objects in the scene.

Fig. 1 A sample frame from long range imagery is shown on the *left* and the objects to be tracked are shown on the *right*

This chapter is organized as follows. The second section explains in detail the preprocessing steps like image registration and moving object detection that are employed to facilitate the tracking procedure. The third section explains in detail the development of the framework for tracking. The results are presented and analyzed in the fourth section followed by a fifth section containing conclusions and a brief preview of the current research that is in progress as future work of the presented results.

2 Feature Extraction Methodology

The key idea in the development of a feature extraction methodology for long range video is to utilize all the available information to aid in the implementation of a robust tracking mechanism. In recent years, the representation of gradient

Fig. 2 Harris corners detected on the objects in wide area imagery. Notice that the number of points (in *red*) detected is very less. It would be hard to depend on very few points to track such an object for a long period of time

information as a histogram has become a very popular technique to represent objects. Scale Invariant Feature Transform (SIFT) [2] descriptors use the information of the gradients in the neighboring cells and this idea was generalized and used for pedestrian detection by Dalal and Triggs [9].

In our research, we exploit a similar idea to represent all the pixels that represent an object. Since we could have a dense representation of the object by employing such a technique, we are able to track many features in successive frames. Similar approach has been employed before to extract the correspondence between neighboring frames with significant difference in content [10].

2.1 Implementation Details

In the original implementation of histogram of oriented gradients, scale was considered in the computation process. However, we do not use scale due to the lack of variation in scale of the object of interest in successive frames. The initial step in the computation is to evaluate the gradient image. We use a simple edge operator employing the following two filters: $H_x = [-1 \quad 0 \quad 1]$ and $H_y = [-1 \quad 0 \quad 1]^T$ on the image I. The resultant gradients denoted as G_x and G_y are then used to find the magnitude $\left(G_{mag}\right)$ and orientation $\left(G_{ang}\right)$ of the gradients. The computational steps for obtaining the gradient information are described in the following Eqs. (1–4).

$$G_x = H_x \otimes I \tag{1}$$

$$G_y = H_y \otimes I \tag{2}$$

$$G_{mag} = \sqrt{G_x^2 + G_y^2} \tag{3}$$

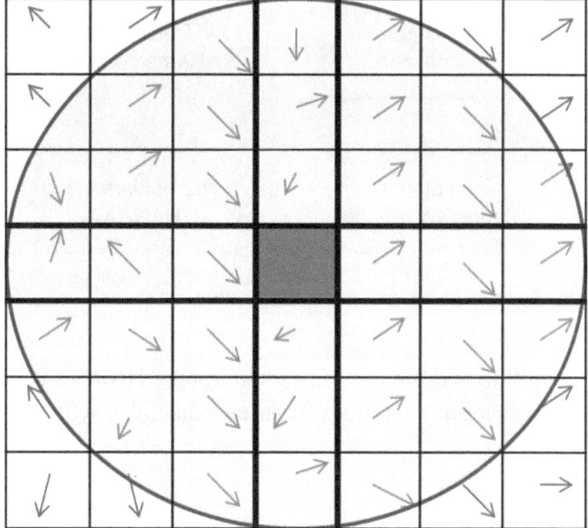

Fig. 3 Illustration of construction of histogram of gradients. The arrows indicate the orientation of the gradient and length of the *arrow* is a measure of the gradient magnitude

$$G_{ang} = tan^{-1}\left(\frac{G_y}{G_x}\right) \tag{4}$$

For every pixel for which the feature vector has to be computed, a local neighborhood patch is defined. The neighborhood patch is divided into cells of a fixed size. A weighted histogram for each neighboring cell is calculated based on the orientations and the weights are decided by the magnitude of the gradient. Multiple histograms from various cells are concatenated to get a highly distinctive descriptor. We have considered a 16×16 patch as the neighborhood. There were 4×4 cells and eight bins for the histogram. This configuration was similar to the original implementation of SIFT [2]. This results in a $4 \times 4 \times 8 = 128$ element vector for each point. The process for binning the orientation for one cell can be expressed as in (5).

$$f(C_x, C_y, b) = \sum\sum I\left\{\left\lceil\frac{G_{ang}(x,y)}{\pi/8} = b\right\rceil\right\} \times G_{mag}(x,y) \tag{5}$$

where (x, y) is the location of the point, C_x and C_y denote the cells, b is the index of the bin and I is the image. The histograms obtained for the various cells can be concatenated to obtain the final feature vector. An illustration of this process is given in Fig. 3.

The main advantage of using such a feature description is the higher dimensionality. Higher dimensionality corresponds to better discrimination power during feature matching. The use of gradients for representing the object provides the best option for describing the shape of the object.

This section has explained the extraction of features from a point set in an image. The assumption at this stage is that the original image on which feature

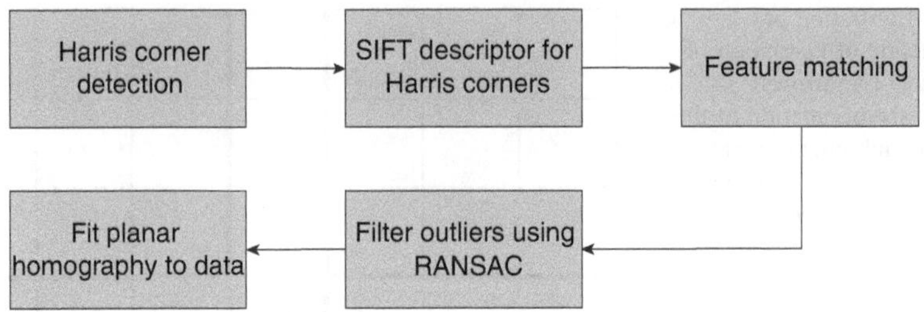

Fig. 4 Framework for image registration

extraction is done is a gray scale image. However, the gray scale image may not be the best input to extract features especially when it comes to tracking very small objects.

3 Image Registration and Moving Object Detection

The pre-processing steps involved in the implementation of the tracking algorithm are image registration and moving object detection. We follow the procedure presented in [11] for these pre-processing steps.

3.1 Image Registration

In order to detect changes in the frames of video, it is essential that the relationship between frames be understood. This process of registering images is especially critical in the case of wide area imagery because of the large shifts in frames. We adopted the framework from [11] as shown in Fig. 4 to register frames. Researchers have been able to register frames from the CLIF database using a correlation based approach [12], robust data alignment [13], by matching SIFT [2] feature points [12] and using Lucas–Kanade algorithm [14, 15]. There is a comparison of the performance of these algorithms in [16]. We made an attempt at using phase correlation [17] to register images in the CLIF dataset. The method failed when there were large translational or rotational shifts in the frames.

The registration method starts with the detection of Harris corners [3]. A point in an image is detected as a corner based on a measure given in (6).

$$A = \sum_{x,y} w(x,y) \begin{bmatrix} I_x^2 & I_x I_y \\ I_x I_y & I_y^2 \end{bmatrix} \tag{6}$$

Fig. 5 Two frames after registration and stabilization

where $w(x, y)$ is the window around the pixel (x, y) and $I_x = \frac{\partial I}{\partial x}$, $I_y = \frac{\partial I}{\partial y}$. When the corner measure given by A in (6) is above a particular threshold, the point is said to be a corner. The points that have been detected as Harris corners are then represented using SIFT feature descriptors. This SIFT feature descriptor is a histogram of gradients in the neighborhood of an interest point. It is computed as the weighted histogram of the orientations of the surrounding points where the weights are the magnitudes of the gradients. The matching criterion for the feature descriptors is as described in [2]. The ratio of the distances of the feature vector from the first and second closest matches are computed. A match is made only if the ratio is greater than a pre-defined threshold. The outliers are filtered using Random Sample Consensus (RANSAC) algorithm [18]. If a projective transformation is assumed, a planar homography matrix (H) can be estimated from the matched points using the equation in (7).

$$P_2 = HP_1 \tag{7}$$

where P_1 is the set of points in the preceding frame, and P_2 is the set of points in the current frame. Although this method is able to register planar surfaces correctly (as shown in Fig. 5), parallax errors occur when there are tall structures in the frame. This is because of the fact that we assume a planar homography for the transformation.

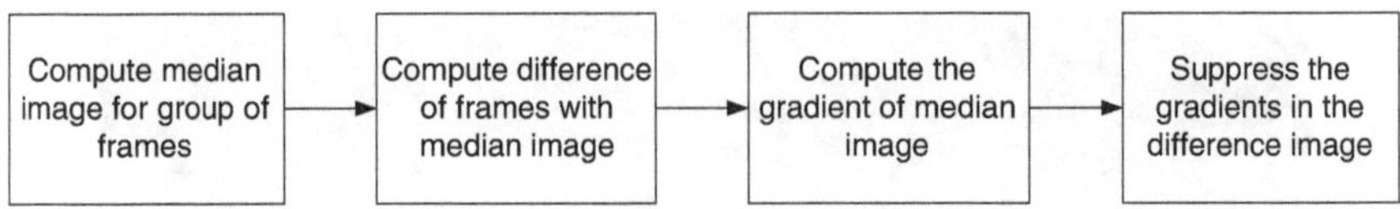

Fig. 6 Method to detect moving objects in a frame

Fig. 7 The original image is shown on the *left* side and the corresponding binary image containing only the moving objects is shown on the *right* side

3.2 Moving Object Detection

It is observed that most of the objects of interest that are to be tracked are objects which change position from time to time. This necessitates the need to have a moving object detection framework. This method is adopted from [11] and the steps are as shown in Fig. 6.

The detection method starts with the computation of the background model for the frames. The background is computed as the median image of a particular number of successive frames. The median image is subtracted from the original frame to obtain the moving objects in the scene. As mentioned in the previous section, the resultant image after registration contains parallax errors and these errors show up as false detections. It is observed that the false detections happen along the edges. In order to eliminate the false detections, we apply a gradient suppression technique. The gradient of the median image is calculated and it is subtracted from the image containing moving objects. The result of moving object detection is illustrated in Fig. 7.

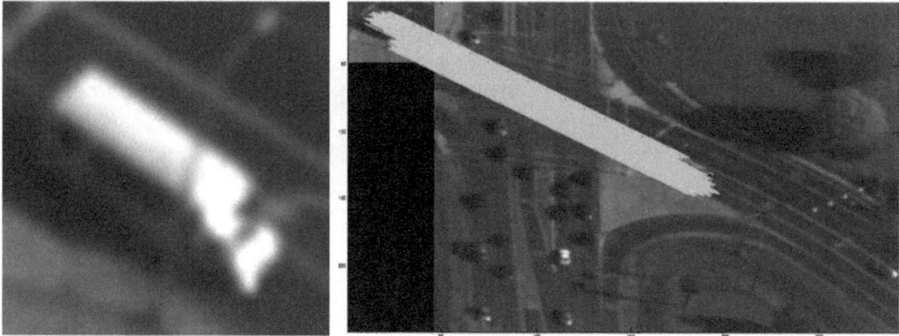

Fig. 8 Illustration of feature matching using a dense set of features. On the *left* side, the object whose features are matched and on the *right* side, the features of the object matched to the object in the scene

4 Proposed Framework for Tracking

This section explains in detail about the framework that has been developed for tracking objects in long range video. In the first few sections described here, the algorithm for moving object detection is not used. The framework is developed based on the process of detecting the object of interest in subsequent frames by matching features after the object is marked by the user.

It is observed that the intensity features are the best available information in the frames for tracking objects of interest on wide area motion imagery. The objective is to compensate for other changes that can happen to the appearance of the object and then devise a robust tracking mechanism. The aim is to have a representation mechanism for a few pixels that are available to represent the object. This was the motivation to develop a region based tracking mechanism. As there are very few pixels to represent the region, we used the method of histogram of oriented gradients on each pixel along with its neighborhood to have a solid feature representation for each object. By having all the pixels that form the object contributing towards the representation model that describes the object, we do not lose any information. The matching of dense set of features is illustrated in Fig. 8. This dense representation of the object is used in the following sections where we evolve a new framework for tracking in long range videos.

4.1 Preliminary Framework for Tracking

The preliminary framework for tracking is shown in Fig. 9. This framework is developed from the basic methodology of detecting an object to be tracked in subsequent frames based on its features.

In Fig. 9 framework starts with representation of the object of interest using a dense feature set. A search area is defined in the next frame where the object is likely to be present. The feature set is extracted for the search area. The features of

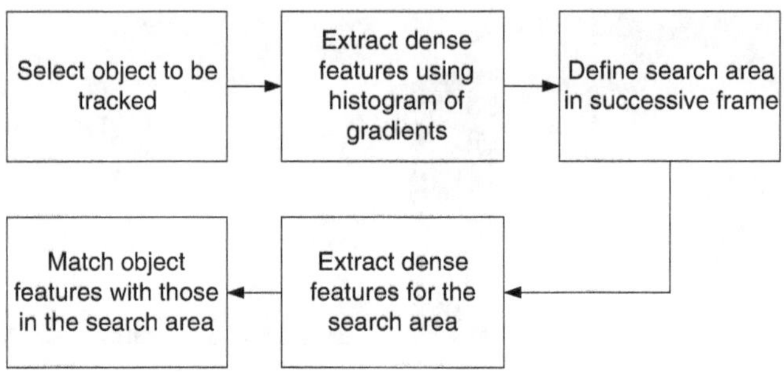

Fig. 9 Preliminary framework for tracking

Fig. 10 The tracker loses the object when it enters the shadow. On the *left*: the vehicle is marked in *red* before entering *shadow region*. On the *right*: the *red box* is the output of the tracker which is not on the object to be tracked (shown in *yellow*)

the object are matched to the features of the search area. The matching criterion is the same as that used in [2]. The method worked well when there was not much variation in the illumination in the scene. However the tracker loses the object when illumination changes as shown in Fig. 10.

4.2 Framework for Tracking in Shadows

The basic idea behind tracking in shadows is to enhance the local region where tracking is being performed. There are a host of image enhancement methods like adaptive histogram equalization [19] and multi-scale retinex [20]. We wanted a method that could best enhance the gradient information in the image to improve the tracking process. The method in [21] was found to be effective in our case. The aim is to have the object region and the search region to have identical illumination

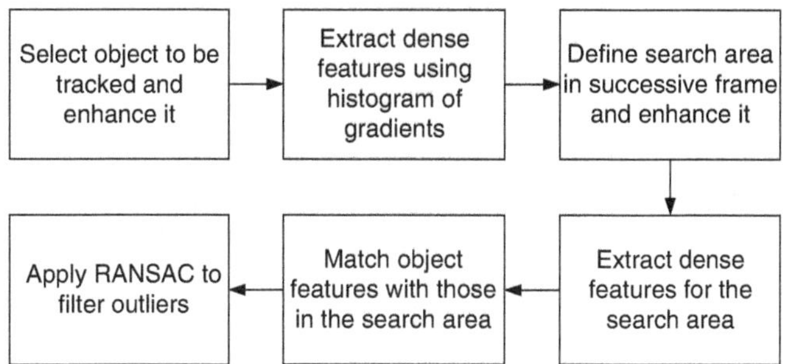

Fig. 11 Framework to track objects in shadows

levels so that the features are matched correctly. The enhancement technique in [21] is non-linear and its advantage is that while it increases the intensity in dark regions and suppresses the intensity in over-exposed regions. This process is done while the contrast is also enhanced. The basic function used for enhancement is given in (8).

$$I_E(x,y) = sin^2\left(I_n(x,y)^q * \frac{\pi}{2}\right) \tag{8}$$

where $I_E(x,y)$ is the intensity after enhancement, $I_n(x,y)$ is the normalized intensity in the original image, and q is a parameter which is estimated adaptively from the local mean of the pixel region.

The modified framework with image enhancement is shown in Fig. 11. The framework has been modified to include the enhancement procedure as well. The object is enhanced as well as the search region. The method fails to work if either one of the regions is not enhanced. The matching of features is done using the nearest neighbor algorithm. This is motivated by the psychological perspective of human visual system. When we try to track objects using our eyes and the illumination decreases, we relax our constraints and come up with a probabilistic measure as to where the object is mostly likely to be at. Similarly, the nearest neighbor approach relaxes the feature matching criterion and also introduces a probability measure to the matching procedure. A proper match is said to be made only if there is a minimum number of matches are made between regions. The threshold is defined based on the size of the object that is being tracked.

The effect of image enhancement on tracking is illustrated in Fig. 12.

4.3 Effect of Super-Resolution on Tracking

We have developed our tracking methodology based on robust matching of features across frames. The robustness of the tracking procedure is critically dependant on

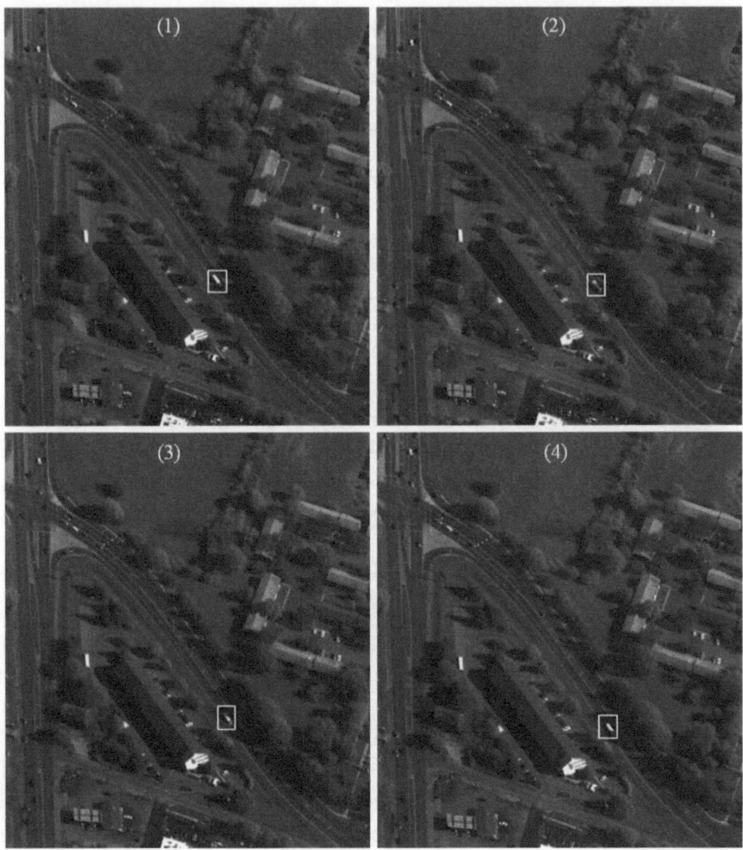

Fig. 12 The object is tracked continuously through frames *1*, *2*, *3* and *4* even in the presence of shadows

the density of features that can be used for matching. Super-resolution is a technique where the number of pixels used to represent an object can be increased and thus provides a denser set of features. The effect of super-resolution becomes evident when the object being tracked is partially occluded. The number of features that represent the region that is not occluded is increased and thus the probability of finding a denser set of matches increases. The advantage of super-resolution in the framework of tracking is illustrated in Fig. 13. Figure 14 illustrates a scenario where an object is tracked continuously even when it is partially occluded.

A lot of research has gone into the development of algorithms for super-resolution [22]. But most of the algorithms depend on the information of multiple frames of the same scene to reconstruct scene information. However, these methods will not be suitable in this case because of the very low frame rate of the video. Therefore we had to increase the resolution based on the information from a single image and the technique in [23] was adopted to do the same. In [23], a

Fig. 13 Tracking partially occluded objects with the aid of super-resolution. **a** The object of interest (car) is marked in *yellow*. **b** The car is partially occluded by a tree. **c** The original car being tracked is matched to the car behind it in the frames with original resolution. **d** The car being tracked is matched to the correct one in the next frame even when partially occluded after the resolution is increased

Fig. 14 Result of tracking with the aid of super-resolution and image enhancement

regression kernel is learned from the phase information present in the local neighborhood. This learned model will be able to adaptively estimate the intensities in the high resolution image.

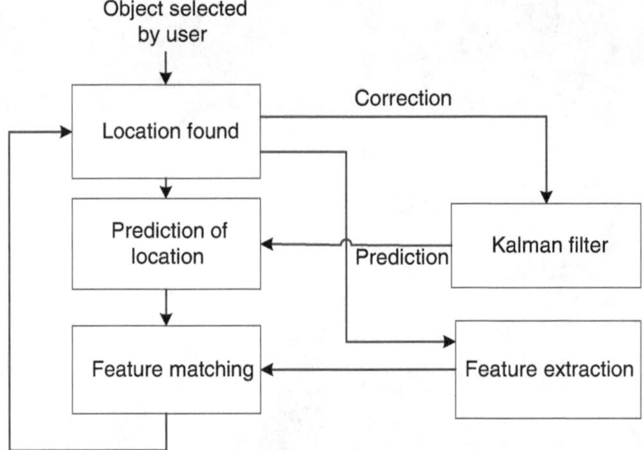

Fig. 15 Framework for the prediction mechanism

4.4 Framework for Tracking in the Presence of Occlusions

A major challenge that needs to be overcome while attempting to track objects in long range videos is the presence of occlusions. In our approach, we have used the Kalman filter [24] as the predictive mechanism. The filter learns the pattern based on the previous information and makes a prediction. If the object is detected in the current frame, the tracker is corrected according to the available information. If the object is not detected in the current frame, the output of the Kalman filter is considered to be the expected location of the object. The framework that illustrates the predictive mechanism is shown in Fig. 15. The result of the tracking process is shown in Fig. 16.

4.5 Framework for Tracking Very Small Objects

So far, we discussed about the development of the tracking algorithm in terms of detecting and tracking objects like cars and trucks. The problem of tracking vehicles has been attempted before by other researchers too [11]. However, there has been no mention of tracking very small objects like pedestrians in long range videos like the CLIF data. In this section, we present the modification of the framework that we have used for tracking vehicles to make it suitable for tracking very small objects.

The main constraint in tracking objects of very small size in such long range videos is their low resolution. While the cars that were being tracked till now had at least 100 pixels, a pedestrian will have only a size of 10 to 15 pixels. Therefore,

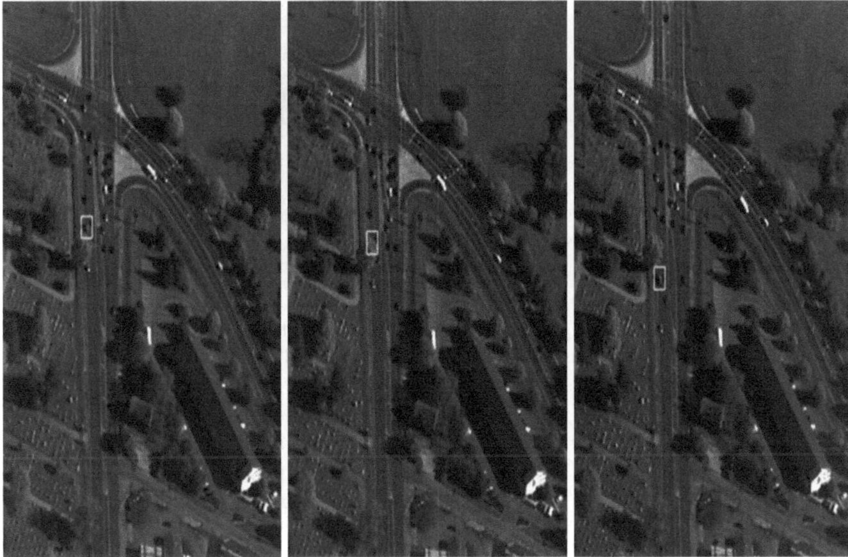

Fig. 16 Result of tracking objects when occluded. Note that the car that passes beneath a tree in successive frames is tracked accurately

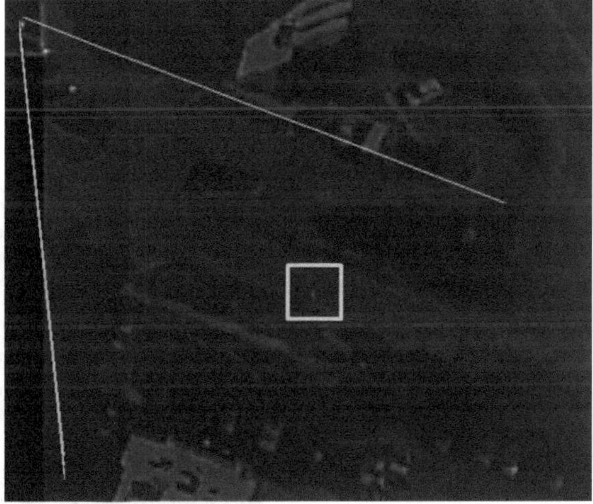

Fig. 17 The pedestrian in the inset needs to be matched to the one in the *yellow box*. However, the matches are not made using the current framework

when the framework mentioned in the previous section is applied, it will not be possible to track small objects like pedestrians. This is illustrated in Fig. 17.

We considered two options to be included into the tracking framework to help enable tracking of pedestrians. The first option is to consider the use of an online learning methodology similar to that presented in [25, 26]. This method tracks an object using 2-bit binary pattern patches, learns the shape manifold of the object

Fig. 18 The pedestrian in the figure on *left* should be matched to the one inside the *yellow box*. However it is matched elsewhere when matching is done based on Gabor filter responses

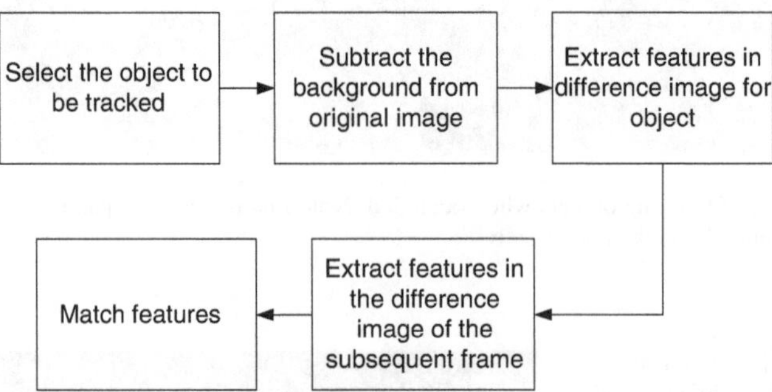

Fig. 19 Framework to track very small objects in long range imagery

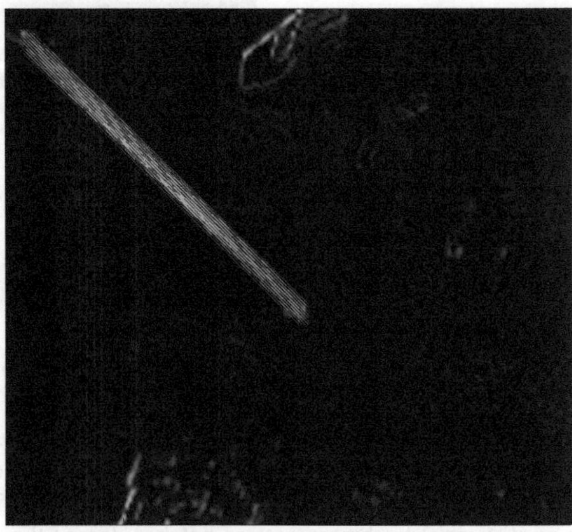

Fig. 20 The features being matched in difference images

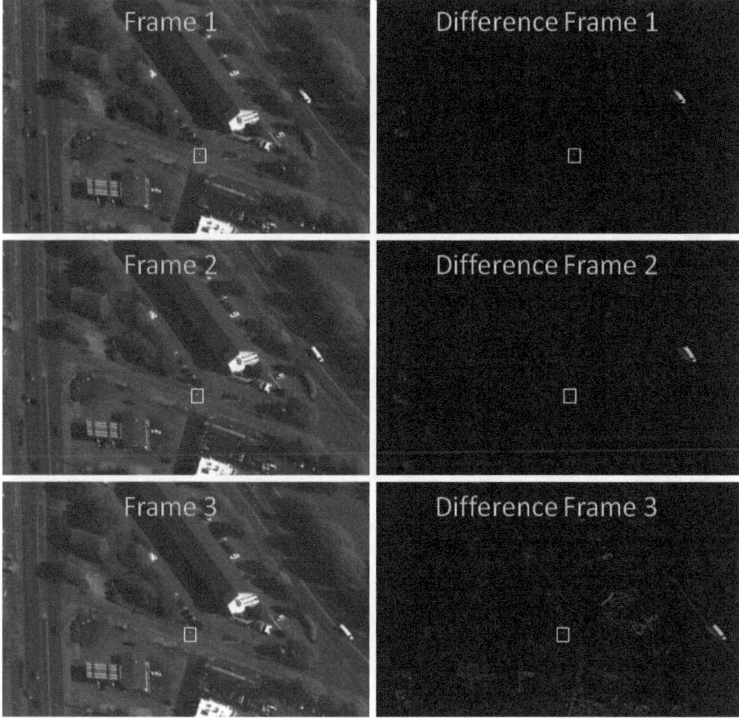

Fig. 21 Result of tracking a very small object (inside the *red box*). On the *left* is the result of the tracker on the original image. On the *right* is the result of the tracker on the images where the background is absent

being tracked online and detects it in subsequent frames using a binary classifier. The method will work well when there is sufficient resolution for the object being tracked. In our case, we would have required a very strong classifier.

The second option is to have feature vectors of higher dimension based on filter responses used for textural analysis. We experimented with this option using Gabor filters [27]. The Gabor filter responses in different orientations are computed. The responses are then encoded into a histogram. The histograms are matched to track the pedestrian. However, the method failed to track the pedestrian. This is because although the dimension of the feature vector representing the object of interest increased considerably, there is no additional information added to the feature representation. The result of matching by this procedure is shown in Fig. 18.

The main observation from the failed attempts so far is that the pedestrian features are matched to some other stationary objects in the scene. This motivated us to have the background removed from the scene. At this stage, we included the moving object detection algorithm into the framework for tracking. The proposed algorithm to track pedestrians is shown in Fig. 19.

The algorithm to model the background was illustrated in Sect. 3.2. Once the background has been subtracted, the remaining elements in the image are the

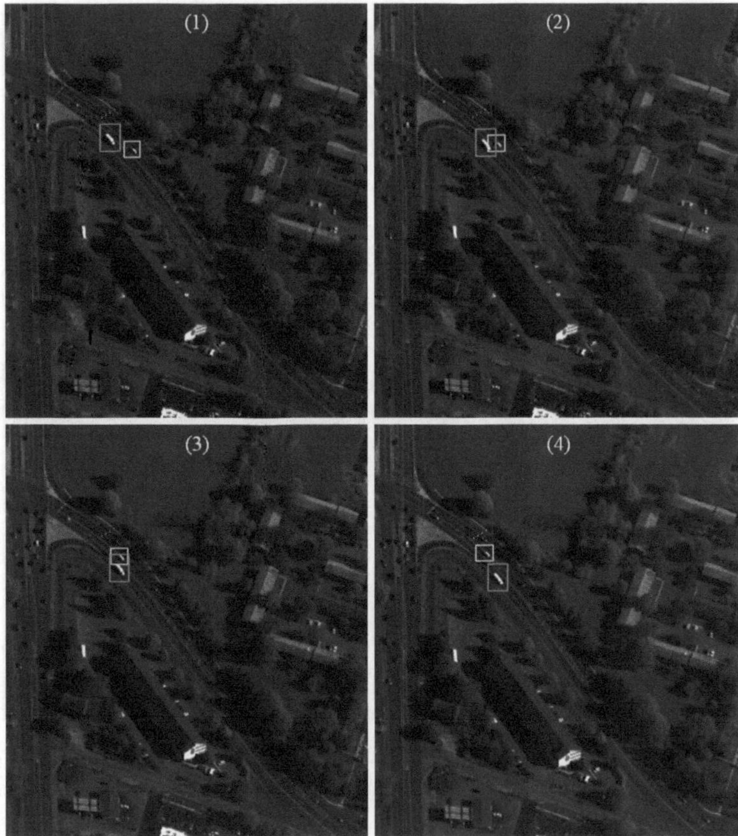

Fig. 22 Multiple object tracking. Note that the vehicles that cross each other are tracked perfectly

moving objects and errors occur due to parallax. When there is sufficient shape or blob appearance of the object in the scene, it would be a significant presence in the resultant image after background subtraction. We have used this fact to track very small objects. The features extracted from the difference image give a very good representation of the object that is in motion. The result of matching a very small object in one frame to the next is illustrated in Fig. 20. The result of tracking very small objects is illustrated in Fig. 21.

5 Results and Discussion

The developed framework has been applied to various videos in varying environments to check the robustness. One of the main areas of concern was whether the tracker would be able to track multiple objects at the same time even when one

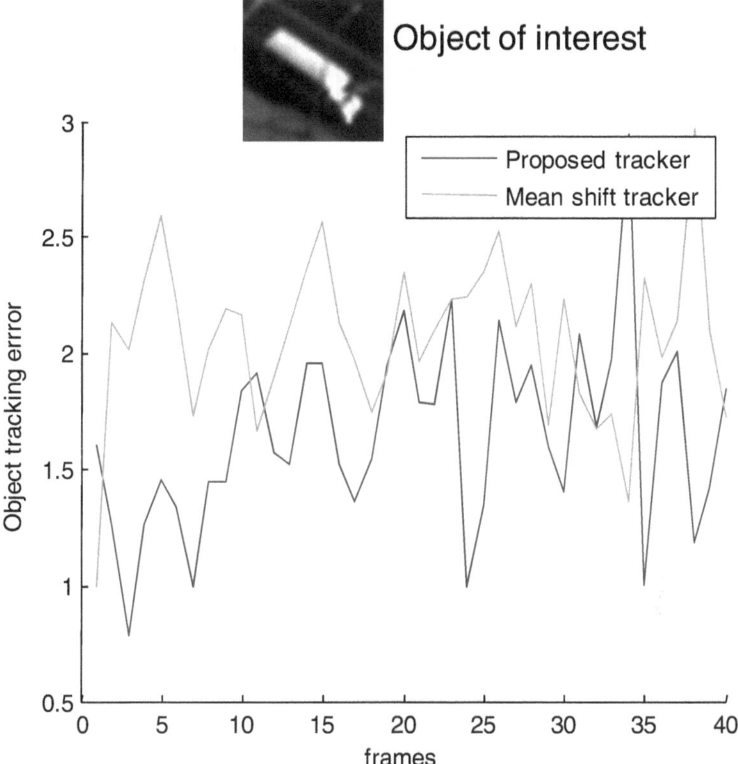

Fig. 23 Comparison with mean-shift tracker to track large objects

object is very near with respect to the other. The result of tracking in such a scenario is illustrated in Fig. 22.

We also did a comparative study with the mean-shift tracker [28] to evaluate the performance of the proposed tracker. Different kinds of objects of interest were tracked to compare the effectiveness. The main focus was on the variation in size of the objects of interest. It was observed that very large objects like trucks can be tracked using the mean-shift tracker as illustrated in the graph shown in Fig. 23. We used a metric called object tracking error as a quantitative metric for comparative evaluation. Object tracking error [29] is the Euclidean distance between the centroid of the bounding box defining the object of interest and the ground truth that is manually marked for each tracked object.

$$\textit{Object tracking error} = \frac{1}{N_{rg}} \sum \sqrt{(xg_i - xr_i)^2 + (yg_i - yr_i)^2} \qquad (9)$$

where N_{rg} is the number of frames for which the ground truth and the detections are available, (xg_i, yg_i) is the ground truth and (xr_i, yr_i) is the detection during tracking.

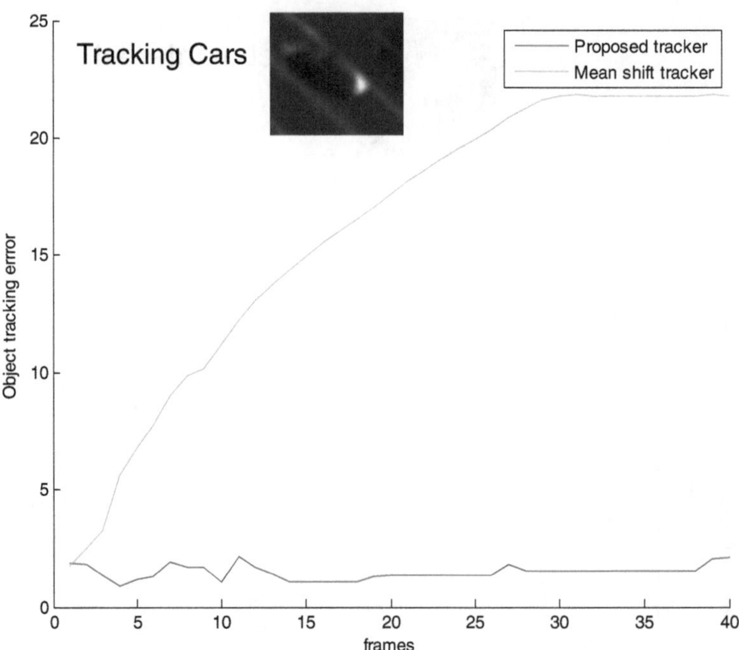

Fig. 24 Comparison with mean shift tracker to track small cars

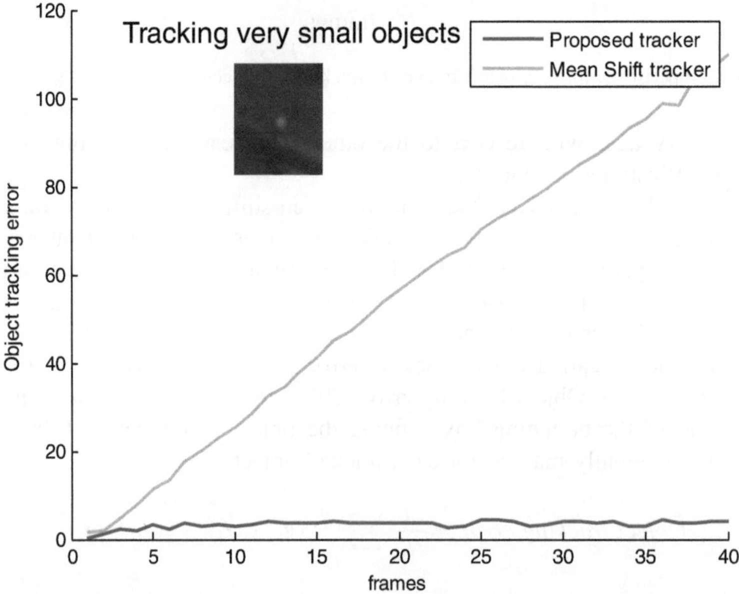

Fig. 25 Comparison with mean shift tracker to track small cars

However, as the size decreased, the proposed tracker worked well while the mean-shift tracker was not successful. This is illustrated in Figs. 24 and 25.

6 Conclusions

The research presented in this chapter is a snapshot of the new technology developed towards tracking extremely low resolution objects in videos with very low frame rate. The proposed method utilized the intensity information available about the object to create a dense feature set based on the histogram of oriented gradients. The framework made use of enhancement and super-resolution techniques to create a better tracking process.

While many of the state of the art techniques concentrate more on learning the object characteristics to create a tracking mechanism, we used features to track the object. Learning models can be incorporated into the proposed framework, where characteristic movement patterns of the object may be learned in a better manner to predict the next location of the object.

The next steps in this research would involve developing models to study the behavior of objects that are being tracked. The development of models that can better describe the movement of the object being tracked is also of importance.

References

1. Yilmaz, A., Javed, O., Shah, M.: Object tracking: a survey. ACM Comput. Surv. **38**, 1 (2006)
2. Lowe, D.: Object recognition from local scale-invariant features. In: International Conference on Computer Vision, Corfu, Greece, pp. 1150–1157 (1999)
3. Derpanis, K.G.: The Harris corner detector. Technical Report, York University. www.cse.yorku.ca/kosta/CompVis_Notes/harris_detector.pdf
4. Mikolajczyk, K., Schmid, C.: An affine invariant interest point detector. In: European Conference on Computer Vision (ECCV). vol. 1, pp. 128–142 (2002)
5. Bay, H., Tuytelaars, T.L., Van Gool, J.: SURF: Speeded up robust features. In: European Conference on Computer Vision (ECCV), pp. 404–417 (2006)
6. Ning, J., Zhang, L., Zhang, D., Wu, C.: Robust object tracking using joint color-texture histogram. Int. J. Pattern Recognit. Artif. Intell. **23**(7), 1245–1263 (2009)
7. Ramanan, D., Forsyth, D.A., Zisserman, A.: Tracking people by learning their appearance. IEEE Trans. Pattern Anal. Mach. Intell. **29**, 65–81 (2007)
8. Columbus Large Image Format Dataset. https://www.sdms.afrl.af.mil/index.php?collection=clif2007 (2007)
9. Dalal, N., Triggs, B.: Histograms of oriented gradients for human detection. In: IEEE Computer Society Conference on Computer Vision and Pattern Recognition, CVPR 2005, vol. 1, pp. 886–893, June 2005
10. Liu, C., Yuen, J., Torralba, A.: SIFT flow: dense correspondence across scenes and its applications. IEEE Trans. Pattern Anal. Mach. Intell. **33**(5), 978–994 (2011)

11. Reilly, V., Idrees, H., Shah, M.: Detection and tracking of large number of targets in wide area surveillance. In: Proceedings of the 11th European Conference on Computer Vision: Part III, Heraklion, Crete, Greece, 5–11 September 2010
12. Mendoza-Schrock, O., Patrick, J., Garing, M.: Exploring image applications registration algorithms for layered sensing. In: O'Donnell, T.H., Blowers, M., Priddy, K.L. (eds.) Evolutionary and Bio-Inspired Computation: Theory and Applications III. Proceedings of the SPIE (2009)
13. Jwa, S., Zhijun Tang, Ozguner, U.: Robust data alignment based on information theory and its applications in road following situation. In: IEEE International Conference on Intelligent Transportation Systems, Toronto, Canada, 2006
14. Lucas, B.D.: Generalized image matching by the method of differences. Ph.D. dissertation, Carnegie Mellon University (1984)
15. Lucas, B.D., Kanade, T.: An iterative image registration technique with an application to stereo vision (IJCAI). In: Proceedings of the 7th International Joint Conference on Artificial Intelligence (IJCAI '81), pp. 674–679, April 1981
16. Mendoza-Schrock, O., Patrick, J.A., Blasch, E.P.: Video image registration evaluation for a layered sensing environment. In: IEEE Aerospace & Electronics Conference (NAECON), 2009 National, pp. 223–230 (2009)
17. Reddy, B.S., Chatterji, B.N.: An FFT-based technique for translation, rotation, and scale-invariant image registration. IEEE Trans. Image Process. **5**(8), 1266–1271 (1996)
18. Fischler, M.A., Bolles, R.C.: Random sample consensus: a paradigm for model fitting with applications to image analysis and automated cartography. Comm. ACM **24**(6), 381–395 (1981)
19. Pizer, S.M., Amburn, E.P., Austin, J.D., Cromartie, R., Geselowitz, A., Greer, T., Romeny, B.H., Zimmerman, J.B., Zuiderveld, K.: Adaptive histogram equalization and its variations. In: Computer Vision, Graphics, and Image Processing, vol. 39, issue 3, 1987, pp. 355–368. Academic Press, San Diego, ISSN 0734-189X, 10.1016/S0734-189X(87)80186-X
20. Rahman, Z., Jobson, D.J., Woodell, G.A.: Multiscale retinex for color rendition and dynamic range compression. In: Tescher, A.G. (ed.) Applications of Digital Image Processing XIX. Proc. SPIE 2847 (1997)
21. Arigela, S., Asari, V.K.: A locally tuned nonlinear technique for color image enhancement. WSEAS Trans. Signal Process. **4**(8), 514–519 (2008)
22. van Ouwerkerk, J.D.: Image super-resolution survey. In: Image and Vision Computing, vol. 24, issue 10, 2006, pp. 1039–1052. Elsevier, ISSN 0262-8856, 10.1016/j.imavis.2006.02.026
23. Islam, M.M., Asari, V.K., Islam, M.N., Karim, M.A.: Super-resolution enhancement technique for low resolution video. IEEE Trans. Consum. Electron. **56**(2), 919–924 (2010)
24. Bishop, G., Welch, G.: An introduction to the Kalman filter. In: SIGGRAPH 2001, Course 8, 2001
25. Kalal, Z., Matas, J., Mikolajczyk, K.: Online learning of robust object detectors during unstable tracking. In: On-line Learning for Computer Vision Workshop, 2009
26. Kalal, Z., Matas, J., Mikolajczyk, K.: P-N learning: bootstrapping binary classifiers by structural constraints. In*: IEEE* Conference on Computer Vision and Pattern Recognition, 2010
27. Movellan, J.: Tutorial on Gabor filters. Technical report, MPLab Tutorials, University of California, San Diego, 2005
28. Comaniciu, D., Ramesh, V.: Mean shift and optimal prediction for efficient object tracking. In: International Conference on Image Processing, 2000
29. Black, J., Ellis, T., Rosin, P.: A novel method for video tracking performance evaluation. In: Joint IEEE International Workshop on Visual Surveillance and Performance Evaluation of Tracking and Surveillance, 2003

Recognizing Complex Human Activities via Crowd Context

Wongun Choi and Silvio Savarese

Abstract This chapter examines the problem of classifying collective human activities from video sequences. A collective activity is defined by the existence of the coherent behavior among individuals in a spatial and temporal neighborhood. Examples of collective activities are queuing in a line or talking. Such types of activities cannot be just defined by considering actions of individuals in isolation but rather by observing the interactions of nearby individuals in time and space. In this chapter we discuss recent methods for analyzing collective activities through the concept of crowd context. We present various solutions for modeling the crowd context and demonstrate the flexibility and scalability of the proposed framework in a number of experiments on publicly available datasets of collective human activities.

1 Introduction

Activities performed by humans or animals have, for the most part, an underlying purpose which characterizes the way individuals interact with each other (Fig. 1). Often activities are defined or reinforced by the existence of a coherent behavior of individuals in time and space, rather than just being based on the behavior of single individuals in isolation. These types of activities are typically called *collective*,

W. Choi (✉) · S. Savarese
Electrical and Computer Engineering, University of Michigan,
Ann Arbor, MI, USA
e-mail: wgchoi@umich.edu

S. Savarese
e-mail: silvio@eecs.umich.edu

Augment Vis Real (2014) 6: 71–91

DOI: 10.1007/8612_2012_4

Published Online: 8 September 2012

Fig. 1 Collective activities are defined or reinforced by the existence of coherent behavior of multiple individuals in time and space. Such coherent behavior is often referred as *crowd context* and it is shown to provide critical cues for understanding collective activities from videos [11, 12]. The figure illustrates examples of collective activities: folk dancing, herd of gazelles traversing the savannah, elephants crossing a pound and football game action

whereas activities that are performed by individuals in isolation are referred as *atomic*. Consider for example the *queuing* activity: the definition of such activity requires that multiple individuals be present in the scene (Fig. 2), standing and waiting their turn in some structure. Over time, the individuals may slowly progress forward in the queue following a coherent spatial and temporal arrangement. This is a collective activity. *Standing* is the activity that each actor is performing in isolation, thus it is an atomic activity. Other examples of collective activities are: people who are *standing* and *talking*; a team of football players performing a game action; herd of gazelles traversing the savannah; a group of elephants crossing a pond (Fig. 1). Notice that the introduction of the concept of collective activity enables the discrimination of activities that are inherently ambiguous if one looks at the individual in isolation (Fig. 2). As figure shows, by just looking at the single individuals it is very hard to tell whether they are in a queue or they are talking (they all appear to be in a standing pose). However, the contextual information provided by the configuration of neighboring individuals can help discriminate that some individuals are standing in a queue and others are standing and talking.

In this book chapter we review a recently proposed framework for representing and modeling collective activities from video sequences [11, 12]. This framework differs from previous research on action classification wherein activities are analyzed by considering individuals in isolation (atomic activities) rather than their collective contextual behavior. In the case of collective activities, action classification may be enhanced by taking advantage of the contextual information that originates from the dynamical (location, velocity, acceleration), geometrical (object pose) and semantic (activity labels) properties of multiple individuals in

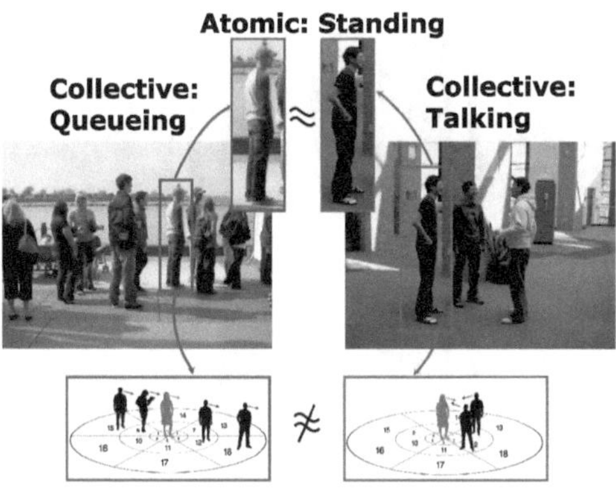

Fig. 2 As introduced in Choi et al. [11, 12], by studying the coherent behavior of individuals in time and space (crowd context) it is possible to effectively characterize human interactions and discriminate collective activities. In isolation, the highlighted individuals have very similar appearance and thus it is not possible to identify whether they are talking (*red*) or standing in queue (*blue*). However, by considering the spatio-temporal distribution of others (i.e. the crowd context) it becomes easier to recognize that the two individuals are performing different activities and to identify which activities are being performed. The spatio-temporal distribution of people, relative to an anchor person, is illustrated in the *lower* part of the figure, where individuals are bucketized over a radial support, based on their physical location in the scene

the surrounding area. This contextual information is often referred as *crowd context* [11, 12]. Moreover, unlike many previous or state-of-the art contributions in human action classification, this study focuses on methods that work under unrestrictive conditions such as moving cameras and cluttered background as well as are robust to variations in illumination and viewpoint conditions. As previous research has demonstrated, this can be achieved by employing methods that are capable of tracking multiple targets in the 3D physical space while estimating the camera motion [9, 10].

One of the objectives of this chapter is to describe different strategies for capturing the crowd context. The original concept of crowd context is introduced in [11] and further extended in [12], where a descriptor is introduced to capture the coherent behavior around each individual. The main idea is to adaptively bin the spatial-temporal volume around a reference individual using a novel classification scheme to infer the most discriminating spatio-temporal volume over which to calculate the crowd context. The proposed adaptive binning strategy has the advantage of: (1) establishing robustness to clutter, (2) being able to incorporate other cues/evidence gracefully for classification, and (3) exhibiting parameter free learning under a principled probabilistic framework. The proposed framework is experimentally evaluated using the dataset introduced in [11] (see Fig. 3). Several experiments are presented to quantitatively and qualitatively assess the ability to classify collective activities via crowd context.

Recognizing collective activities is valuable in numerous video surveillance scenarios wherein it is key to track individuals, interpret and describe their

Fig. 3 Frames extracted from the dataset on collective activities introduced in [12]. The dataset comprises video sequences that contain six different collective activities with a variable number of actors in the scene: *Crossing, Waiting, Queueing, Talking, Dancing* and *Jogging*

behavior at various degrees of semantic resolution. Moreover, modeling collective activities plays a critical role in related research areas such as robotics and autonomous navigation as well as in applications where the content in large video repositories must be indexed, searched and organized. Finally, it provides tools for analyzing and studying typical or anomalous spatial-temporal collective behaviors in biology (insects, animals) or biomedicine (cells) and help construct an ontology of human or animal of complex behaviors.

The rest of this chapter is organized as follows. In Sect. 2, we review previous works on activity classification from videos. In Sect. 3, we present state-of-the-art methods for estimating spatial-temporal trajectories of multiple targets from moving cameras. In Sects. 4 and 5, we discuss a recent framework that use these trajectories to capture the crowd context and use it to classify collective activities. An overview of main experimental results are presented in Sect. 6 and conclusions are drawn in Sect. 7.

2 Related Works

A large literature on activity classification has mostly focused on understanding the behavior of humans in isolation (*atomic activities*). Song et al. [46] and Fanti et al. [17] model actions using a constellation of parts and relate the spatial-temporal dependencies of such parts using a probabilistic formulation. Laptev and Lindberg [27] and Dollar et al. [14] propose a compact characterization of an activity in terms of a sparse set of local spatial-temporal interest points. Savarese et al. [45] introduce a framework for incorporating 2D spatial-temporal short and long term dependencies into a bag-of-words representation. Niebles et al. [39] introduce an unsupervised learning method for modeling activities that leverage the construction of latent intermediate visual concepts. Other interesting formulations can be found in the works by [23, 31, 32, 34–37, 52, 55] and are nicely summarized in the survey by Turaga et al. [48]. Progress on atomic activity recognition is coupled with the effort of collecting datasets of human activities that appear in images or videos. Early notable examples are the *KTH* dataset [19] and the *Wisemann* dataset [19]. More recent collections are proposed by Liu et al. [33] (a data set of videos from a public video repository (YouTube) and Laptev et al. [28] (a collection of video sequences from Hollywood movies) which provide a test-bed that is closer to real world application scenarios. Recently, Niebles and Fei-Fei [38] propose to model structured atomic activities such as those that appear in sport events (e.g. *tennis-serve*, *triple-jump*, etc.) and provide a large dataset for enabling quantitative evaluation.

While successful, however, most of these methods are targeted to atomic activities. Research by Ryoo and Aggarwal [43], Yao et al. [54] and Patron et al. [40] goes beyond single-person activity understanding and propose methods for modeling interactions between pairs of individuals. The extension to activities that involve more than two individuals has been investigated in a number of works including [2, 25, 44]. In Ryoo and Aggarwal [44] complex group activities are analyzed using a stochastic context free grammar model with a number of predefined activity predicates. Lan et al. [25, 26] propose to encode 2D interactions among individuals using the contextual information that originates from higher level activity semantics. Choi et al. [11, 12] focus on activities that are characterized by a larger number of individuals (e.g. the collective activities) and propose to capture the collective behavior using a descriptor called *Crowd Context*. Choi et al. [11, 12] also propose one of the first data sets that include challenging videos of collective activities. We shall discuss this method in details in Sect. 6. Moreover, a number of works [21, 30, 47] focus on group activities that appears in sport events such as a football game. Specifically, Intille and Bobick [21] model trajectories of individuals with a Bayesian network, Li et al. [30] introduce a discriminative temporal interaction manifold for modeling activities [30] and Swears and Hoogs [47] propose a non-stationary kernel hidden Markov model to capture temporal dependencies. Notice that most of these methods require different degrees of manual annotations in identifying human trajectories in time and space. Finally, at the opposite side of the spectrum, research by [20, 42, 57] seeks to study the semantic properties of large crowds of individuals. These methods, however, go beyond the scope of this chapter in that the focus is on modeling large crowds as a whole, without considering the individual behavior of the actors.

3 Localizing People in 3D Space from a Video Sequence

As we will describe in more details in Sect. 4, the crowd context descriptor [12] captures the coherent behavior of individuals in a certain spatial-temporal neighborhood. Such coherent behavior can be modeled by estimating the spatial-temporal trajectories of individuals that are performing a certain activity in the scene. These can be, in turn, determined by using tracking algorithms that are capable of simultaneously identifying the spatial location of multiple targets in each frame of the video sequence. These methods are often referred as *multi-target tracking* algorithms. In the next sections, we briefly review recent methods for multi-target tracking and focus on the method proposed by Choi and Savarese [10] (Sect. 3.1). The approach has the ability to estimate the spatial-temporal trajectories in the 3D physical space and, thus, enables the construction of a crowd context descriptor that is invariant to view point transformations and robust to occlusions.

By leveraging the recent development of techniques for accurate and efficient object detection [13, 18, 29, 49], a new paradigm for object tracking called *tracking-by-detection* is explored in the works by [3, 7, 10, 22, 41, 51, 53, 56]. For instance, Wu et al. [23] and Breitenstein et al. [7] propose a system that can track multiple human targets by estimating most probable target's location using detection responses. Zhang et al. [56] and Pirsiavash et al. [41] treat the tracking problem as a data association problem using a linear or dynamic programming formulation. All of these techniques focus on localizing targets in the 2D image plane.

Recently, research by [9, 10, 15, 50] have explored methods for multi-target tracking from moving cameras. Wojek et al. propose a probabilistic framework to detect individuals by combining multiple detectors [51] and by explicitly reasoning about target self-occlusions [50]. Ess et al. [15, 16] propose a method that leverages camera odometry to obtain more robust tracking results. As discussed in details in Sect. 3.1, Choi et al. [9, 10] introduce a technique for estimating camera motion and tracking multiple people in a coherent framework from un-calibrated monocular video sequences. Unlike the aforementioned methods, Choi et al. [9, 10] do not require that information about the camera odometry is available.

3.1 Model Overview of the Tracking System

In this section, we review the tracking method by Choi et al. [10] which allows to estimate the 3D trajectories of an unknown number of targets given a video sequence recorded from a monocular moving camera. Suppose an input video sequence $\mathbb{V}$ is available and that $\mathbb{V}$ is composed of a number of frames at different time stamps ($\mathbb{V} = \{I_0, I_1, \ldots, I_T\}$). Then, the problem of multi-target tracking can be formulated as a sequential Bayesian inference problem that seeks to find the maximum-a-posteri (MAP) solution of the following probability:

$$P(\Omega_t|I_{0,\ldots,t}) \propto \underbrace{P(I_t|\Omega_t)}_{(a)} \int \underbrace{P(\Omega_t|\Omega_{t-1})}_{(b)} \underbrace{P(\Omega_{t-1}|I_{0,\ldots,t-1})}_{(c)} d\Omega_{t-1} \tag{1}$$

where Ω_t is the *configuration* variable that includes the camera parameters Θ_t, all targets' states Z_t in 3D and all image features' states G_t in 3D. Thus, the problem of multiple target tracking is formulated as finding the most likely configuration Ω_t

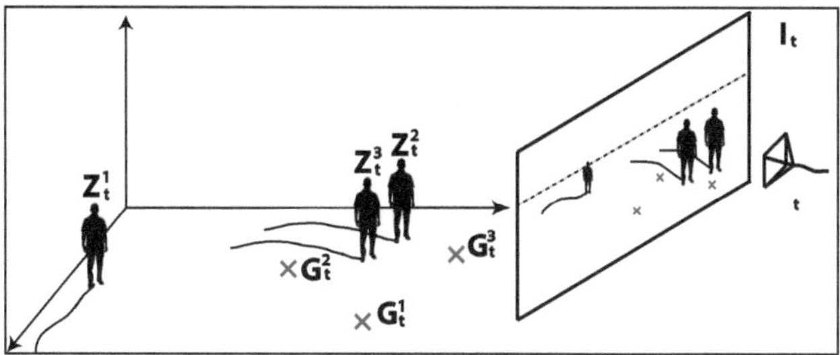

Fig. 4 Given a video sequence $\mathbb{V} = \{I_0, I_1, \ldots, I_T\}$, the goal of multiple target tracking is to find the configuration $\Omega_0, \Omega_1, \ldots, \Omega_t$ that is most conforming with the observed data $\mathbb{V}$. The configuration variable Ω_t is composed of the camera parameters Θ_t, the set of target's location in 3D space Z_t and the set of geometric features' states in 3D space G_t. A simplified camera projection function can be utilized in order to define the transformation of 3D targets Z_t and features G_t onto the 2D image plane given camera parameters Θ_t

given input observations $I_0, I_1, \ldots, I_t$ up to time-stamp t. The geometric features are adopted into the system in order to provide a more robust and accurate estimation of the camera motion parameters.

The first term (**a**) in the Eq. 1 represents the observation likelihood of the configuration Ω_t given the image I_t at time-stamp t. For every possible configurations of Ω_t, one can obtain the image projections of all targets and features as shown in the Fig. 4. The second term (**b**) is the motion model that embeds the smoothness of the camera motion, targets' motion and interaction among targets. Finally, the third term (**c**) is the posterior probability at previous time stamp $t-1$. As Eq. 1 suggests, given the video, observation likelihood and motion model, one can solve the multiple target tracking problem in a sequential fashion; i.e. by assuming that the posterior probability in the 0th frame is available, the posterior probability at arbitrary time stamp t can be obtained by sequentially calculating the posterior probabilities from time 1 to $t-1$.

Assuming that the image observations of each individual target as well as the image features are independent, one can factorize the overall observation likelihood into the product of individual likelihood functions as follows:

$$P(I_t|\Omega_t) = \prod_i P(I_t|Z_t^i, \Theta_t) \prod_j P(I_t|G_t^j, \Theta_t) \tag{2}$$

$$P(I_t|Z_t^i, \Theta_t) = P(I_t|f_{\Theta_t}(Z_t^i)) \tag{3}$$

$$P(I_t|G_t^j, \Theta_t) = P(I_t|f_{\Theta_t}(G_t^j)) \tag{4}$$

where Eqs. 3 and 4 represent individual observation likelihood functions of targets and features, respectively. Any image projection function f_{Θ_t} can be adopted depending on the application, e.g. simplified camera projection model as in [10] or pinhole camera model as in [9]. The likelihood can be estimated either using a combination of various detection responses [9] or by adding Gaussian noise on the detection outputs [10].

Similarly to the observation likelihood, the motion model can be factorized into 3 terms as follows:

$$P(\Omega_t|\Omega_{t-1}) = \underbrace{P(\Theta_t|\Theta_{t-1})}_{(a)}\underbrace{P(Z_t|Z_{t-1})}_{(b)}\underbrace{P(G_t|G_{t-1})}_{(c)} \tag{5}$$

The first term (**a**) in the Eq. 5 can be used to model motion priors on the camera trajectories. A common prior is the smoothness on the camera location parameters. Such motion prior can be expressed, for instance, by a linear motion model, $\Theta_t = A\Theta_{t-1} + V$ where A is a transition matrix and V is a Gaussian random variable that encodes the small uncertainty in the motion of the camera.

The second term (**b**) is the motion model of the targets. Under the assumption that the targets are independent, the motion model can be factorized into the product of individual motion models (Eq. 6).

$$P(Z_t|Z_{t-1}) = \prod_i P(Z_t^i|Z_{t-1}^i) \tag{6}$$

where $P(Z_t^i|Z_{t-1}^i)$ encodes both the linear motion prior (similarly to camera motion model) and the existence prior which captures the property that if a target exists at time frame $t-1$, it is likely to exist at current frame t as well.

Under the assumption that targets are not independent and that there exists an interaction between targets, the motion model of the targets can be represented as in the Eq. 7.

$$P(Z_t|Z_{t-1}) = \prod_{i,j} \Psi(Z_t^i, Z_t^j) \prod_i P(Z_t^i|Z_{t-1}^i) \tag{7}$$

$\Psi(Z_t^i, Z_t^j)$ is a potential function that models the interaction between a pair of people. Examples of interactions can be, for instance, repulsion (targets cannot occupy the same physical space) or attraction (see [10] for details).

Finally, the last term (**c**) in the Eq. 5 is the motion model for the geometric features. The feature motion model is factorized into the product of individual feature motion models each of which encodes priors such as smoothness and existence.

Finding the MAP solution from this probabilistic model is an NP hard problem. Choi et al. [9, 10] employ an approximated sampling scheme to solve the MAP solution in each time frame and propagate the posterior in current time stamp to the next one. The Reversible Jump Markov Chain Monte Carlo (RJ-MCMC) particle filtering is particularly suitable for handling variable number of variables (targets) and complex posterior probability structure. Please refer to the papers [9, 10] for more details. From the MAP, one can easily obtain a set of the spatio-temporal trajectories associated to the targets (individuals) in the scene.

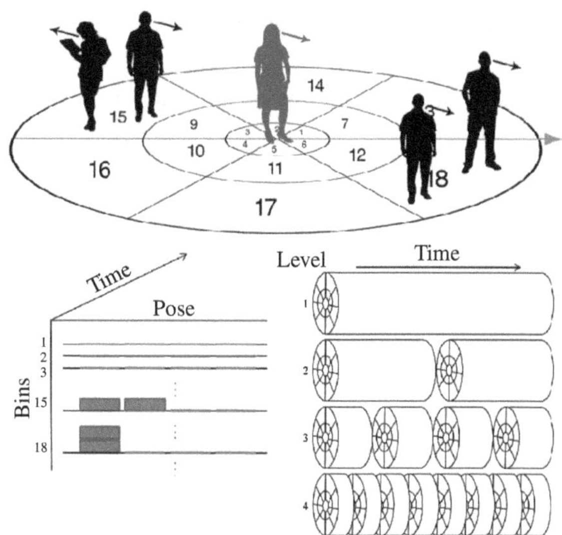

Fig. 5 Spatio-Temporal Local Descriptor. **a** Space around anchor person (*blue*) is divided into multiple bins. The pose of the anchor person (*blue arrow*) locks the "orientation" of the descriptor which induces the location of the reference bin "1". **b** Example of STL descriptor—the descriptor is a histogram capturing people and pose distribution in space and time around the anchor person. **c** Classification of STL descriptor is achieved by decomposing the histogram in different levels along the temporal axis

4 Crowd Context for Collective Activity Recognition

In this section, we introduce the definition of the crowd context and describe its mathematical formulation given a set of spatio-temporal trajectories. The concept of crowd context is first introduced by Choi et al. in [9, 12] and is defined as *a coherent behavior of individuals in time and space* that are performing a certain collective activity.

In [11], crowd context is captured by introducing a new descriptor called Spatio-Temporal-Local (STL) descriptor that encodes the spatial-temporal dependencies of individuals in a neighborhood of the video sequence. The STL descriptor is in essence a fixed-dimensional vector (Fig. 5) and is associated to each person. For each time stamp, the STL descriptors are used to classify the collective activity using a standard Support Vector Machine (SVM) [8] classifier. Temporal smoothness is enforced by applying a markov chain model across each time stamp. Though the method shows promising results, such rigid descriptors require the parameters that control the structure of the descriptor to be manually specified, which can be extremely difficult in presence of large intra-class variability. Such limitation is addressed in [12] where a new scheme called Randomized Spatio Temporal Volume (RSTV) is used to automatically learn the best structure of the descriptor. In following sections, we review the rigid STL descriptor first and the extended RSTV later.

4.1 Rigid STL Descriptor

In this section, we describe how to extract an STL descriptor for each individual (track) in each time stamp given a set of trajectories $\mathbb{T} = \{T_1, T_2, \ldots, T_N\}$, where $T_i = \{l_i, p_i, t_i\}$ is an individual track and $l_i = (x_i, y_i)$, p_i and t_i are sequences of x, y location, pose and time index, respectively. Note that the pose captures the orientation of an individual in this framework (e.g. left, front, right, and back). The location of individual target l_i is obtained by accumulating the estimated state Z_t^i acquired by the multi-target tracking method (as discussed in Sect. 1.3) and pose p_i is acquired by using SVM classifier equipped with HoG descriptor [11].

Given a person i in certain time stamp t (the *anchor*), they determine the locations l_j^i and poses p_j^i of other individuals in the anchor's coordinate system, where the anchor's coordinate system has the origin at the anchor's (x, y) location and is oriented along the pose direction of the anchor (see Fig. 5 top).

The space around each anchor i at time t is divided into multiple bins following a log-polar space partition similar to the shape context descriptor [4]. Moreover, for each spatial bin, P "pose" bins are considered where P is the number of poses that are used to describe a person orientation. Finally, the temporal axis is also decomposed in temporal bins around time stamp t. This spatial, temporal and pose sensitive structure is used to capture the distribution of individuals around the anchor i at time t and construct the STL descriptor. For each anchor i and time stamp t, an STL descriptor is obtained by counting the number of individuals that fall in each bin of the structure described above. Thus, the STL descriptor implicitly embeds the *flow* of people around the anchor over a number of time-stamps. After accumulating the information, the descriptor is normalized by the total number of people that fall in the spatio-temporal extension of the descriptor.

There are a number of important characteristics of the STL descriptor. First, the descriptor is rotation and translation invariant. Since the relative location and pose of individuals are defined in the anchor's coordinate system, the descriptor yields a consistent representation regardless of the orientation and location of the anchor in the world. Moreover, the dimensionality of the descriptor is fixed regardless of the number of individuals that appear in the video sequence. This property is desirable in that it allows to represent an activity using a data structure that is not a function of the specific instantiation of a collective activity. Finally, by discretizing space and time into bins, the STL descriptor enables a classification scheme for collective activities that is robust to variations in the spatio-temporal location of individuals for each class of activity (intra-class variation).

Given a set of STL descriptors (each person in the video is associated to a STL descriptor) along with the associated collective activity labels, one can solve the collective activity classification problem by using a classification method such as SVM [8]. In order to capture various levels of temporal granularity, the authors of [11] adopt SVM classifier equipped with a temporal pyramid intersection kernel (see Fig. 5 bottom right). The temporal axis is divided into 4 hierarchical levels of temporal windows and intersection kernel is defined per each level. The finest

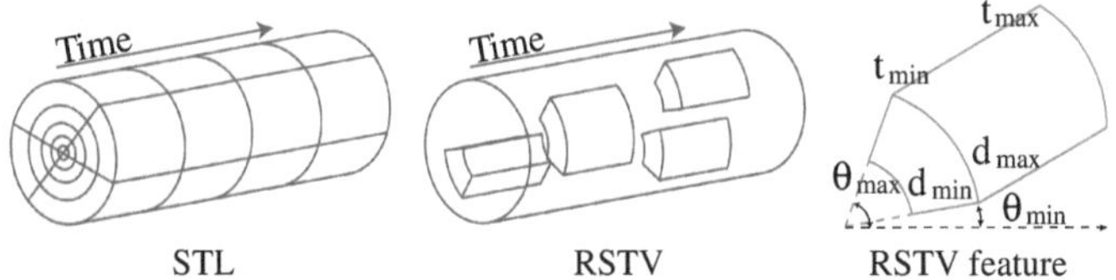

Fig. 6 STL counts the number of people in each spatio-temporal and pose bins that are divided by a hand defined parameters (*left*). On the other hand, the RSTV learns what spatial bins are useful (shown as a trapezoid-like volume) in order to discriminate different collective activities and discards the regions (shown as empty regions) that are not helpful for such discrimination task (*middle*). A random spatio-temporal volume (feature) is specified by a number of parameters (*right*). Pose and velocity are omitted from the illustration

temporal window allows to capture the detailed motion of individuals around the anchors; the highest level allows to encode the overall distribution of people around the anchor over the observed period.

4.2 Learning the Crowd Context

Even though the STL descriptor has been successfully employed for collective activity classification by Choi et al. [11], it is limited in that the structure of the bins of the STL descriptor is predefined beforehand and parameters such as the minimum distance from the anchor or the maximum support of the descriptor are defined once for all. In particular, by assuming that the spatial support has fixed size, the STL descriptor does not have the ability to adaptively filter out background activities or activities that differ from the dominant one.

In order to avoid above mentioned limitations, a novel scheme, called Randomize Spatio-Temporal Volume (RSTV), is proposed by Choi et al. [12]. The RSTV approach is based on the same intuition as STL that crowd context can be captured by counting the number of people with a certain pose and velocity in fixed regions of the scene, relative to an anchor person. However, RSTV extends this intuition and considers variable spatial regions of the scene with a variable temporal support. The full feature space contains the evidence extracted from the entire videos: the location of each individual in anchor's coordinates as well as the velocity and pose of each individual per video frame. This can be interpreted as a soft binning scheme where the size and locations of bins are estimated by a random forest so as to obtain the most discriminative regions in the feature space. Over these regions, the density of individuals is inspected, which can be used for classification. Fig. 6 compares the rigid STL binning scheme and the flexible RSTV. RSTV is a generalization of the STL in that the rigid binning restriction imposed in the STL is removed. Instead, portions of the continuous spatio-temporal volume are sampled at random and the discriminative regions for classification of a certain activity are retained. RSTV provides increasing discrimination power due to increased flexibility.

There are several benefits of the RSTV framework over rigid STL descriptor. (1) The RSTV automatically determines the discriminative features in the feature space that are useful for classification. Indeed, while STL proposes a rigid and arbitrary decomposition of the feature space, in RSTV the binning space is partitioned so as to maximize discrimination power. (2) Unlike STL, there are no parameters that are to be learned or selected empirically (e.g. support distance, number of bins). (3) It enables robustness to clutter. Indeed, unlike STL, the RSTV does not operate given fixed parameters such as radial support and number of spatial bins, but explores the possible space of parameters; thus the density feature, using which classification is performed, is only calculated over regions relevant to each different activity. Hence the classification evidence is pertinent to each activity and avoid clutter that possibly arises from hard-coded framework parameters that may be tuned to achieve optimal classification of a few activities, but not all. Notice that STL concept is similar to the Shape Context [4] descriptor, which is known to be susceptible to clutter due to non discriminative inclusion of all points within the radial support.

Learning RSTV with Random Forest: In [12], Random Forest classifier is used to learn the structure of RSTV given training data. A Random forest [6] is an ensemble of many singular classifiers known as decision trees which is trained from a portion of the training data. The training set is subdivided into multiple *bags* by random sampling with replacement (*bagging*) in order to reduce the effect of over-fitting. Given each set, one random decision tree is trained following successively drawing and selection of a random feature that best discriminates the given training set [6].

The RSTV is trained based on the random forest classifier given a set of training data and associated activity labels (x_i, y_i) where each data point is defined for each person and time stamp. In following description, it is assumed that the trajectories and poses of all people are already transformed into the anchor's coordinate system to form data point x_i and associated activity label y_i. Given a random bag, a random decision tree is learned by recursively discovering the most discriminative features. The algorithm firstly randomizes over different volumes of the feature space and secondly randomizes over different decision thresholds given the feature subspace. The feature is defined as the number of people lying in a spatio-temporal volume that is specified by location (l^k), velocity (v^k), pose (p^k) and time (t) defined in the anchor's (k) coordinate system. A unique spatio-temporal volume is specified by a number of parameters: (1) minimum and maximum distance d_{min}, d_{max}, (2) minimum and maximum angle in the space θ_{min}, θ_{min}, (3) relative orientation/pose p, (4) temporal window t_{min}, t_{max} and (5) minimum and maximum velocity v_{min}, v_{max} (Fig. 6 right). In each node, a number M of such hyper-volume r_n and a scalar decision threshold d_n is drawn randomly multiple times. Given the feature pair (r_n, d_n), the training data is partitioned into two subsets I_r and I_l by testing $f(x; r_n) > d_n$, where $f(x; r_n)$ is a function that counts the number of people lying in the hyper volume r_n. Among the set of candidate features, the one that best

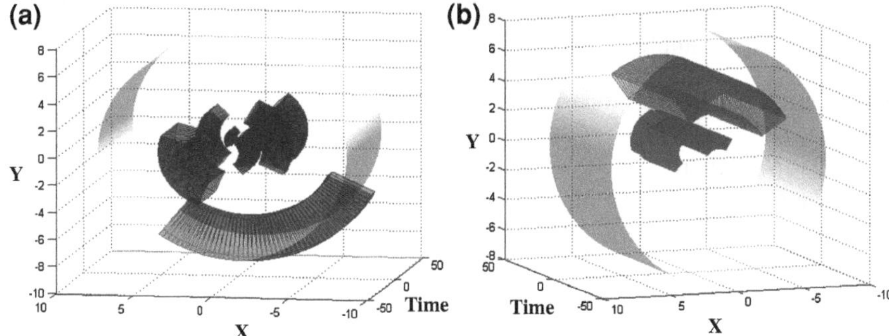

Fig. 7 Example of learned RSTV regions. **a** and **b** illustrate a set of RSTV regions learned automatically by a single tree. Each colour indicates different pose of neighbouring individuals (*up—red*, *down—blue* and *right—green*). Each RSTV is oriented such that the anchor is facing in the upward z direction. Hence **a** indicates that while waiting, an anchor is surrounded on the left and right by people facing the same direction. RSTV in **b** illustrates that during talking the anchor and neighbour face each other and are in very close proximity. Note that each RSTV needs only capture some coherent portion of evidence since there exist many trees in the RF. x and z have units of meters while time is measured in frames

discriminates the training data into two partitions is selected by examining the information gain (Eq. 8).

$$\begin{aligned} \Delta E &= -\frac{|I_l|}{|I|}E(I_l) - \frac{|I_r|}{|I|}E(I_r) \\ E(I) &= -\sum_{i=1}^{C} p_i \, log_2(p_i) \end{aligned} \tag{8}$$

I_l and I_r are the partition of set I divided by given feature, C is the number of activity classes, p_i is the proportion of collective activity class i in set I, and $|I|$ is the size of the set I. Typical examples of learned RSTV structure is shown in Fig. 7. The detailed algorithm for learning RSTV is presented in Algorithms 1 and 2.

Algorithm 1 RSTV learning
Require: $I = \{(x_i, y_i)\}$
  Randomly draw a bag I_t for each tree
  for all random decision tree **do**
    At the root node, $root \leftarrow NodeLearn(I_t)$
  end for

Given the learned RSTV forests, one can classify a novel testing example x by passing down the example along each tree and taking the class that maximizes marginal posterior probability $P(y|x) = \sum_{tree} P_{tree}(y|x)$ over all trees. The posterior probability of a tree is defined as the corresponding p_y in the leaf node that the testing example reached in the decision tree.

Algorithm 2 Recursive Node Learning (NodeLearn)
Require: I_n
  if $|I_n| < N_{min}$ **then**
    Compute distribution of classes p_i over all C
    $node.isleaf \leftarrow TRUE$
    $node.p \leftarrow p_i$
    return *node*
  end if
  $\Delta E_{max} \leftarrow -INF$
  for $m = 0$ **to** M **do**
    Randomly draw a feature pair (r_n^m, d_n^m)
    Compute information gain ΔE_m
    if $\Delta E_{max} < \Delta E_m$ **then**
      $\Delta E_{max} \leftarrow \Delta E_m$
      $(r_n, d_n) \leftarrow (r_n^m, d_n^m)$
    end if
  end for
  Partition I_n into (I_l, I_r) using (r_n, d_n)
  $node.isleaf \leftarrow FALSE$
  $node.left \leftarrow NodeLearn(I_l)$
  $node.right \leftarrow NodeLearn(I_r)$
  $node.feature \leftarrow (r_n, d_n)$
  return *node*

5 Globally Consistent Classification with Markov Random Field

The STL and RSTV models allow to classify each person in the video individually and associate a collective activity label to it. If the scene contains only one or few collective activities, however, one can impose some level of spatial or temporal regularization across labelling assignments. Such regularization helps mitigate the classification errors due to the inherent noise in constructing the STL/RSTV descriptors as well as the intrinsic ambiguities in discriminating collective activities. This regularization is modelled using a Markov Chain as in the work by Choi et al. [11], whereas it is encoded as a Markov Random Field (MRF) over both space and time as in [12]. In this section, we discuss the MRF formulation introduced in [12].

An MRF is a general model that can encode the correlation between many random variables in a coherent fashion. Such model is frequently adopted in the image segmentation problem in order to provide consistency between spatially adjacent pixels [24]. Choi et al [12] propose to use an MRF to capture the local

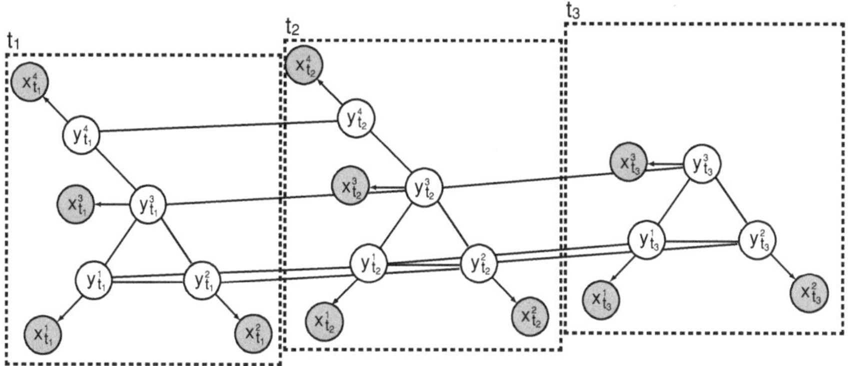

Fig. 8 Graphical representation for the proposed MRF over collective activity variables y. $y^j_{t_i}$ models the activity of a person in one time slice (hidden variable), $x^j_{t_i}$ represents the trajectories associated to an anchor person. If two people are close enough (≤ 2 m away), the spatial edges are inserted to inject spatial coherency. For every person, temporal edges are constructed between nearby nodes

and spatial coherency of labelling assignments. They assign the same activity label to nearby people in a single time stamp and to a person over adjacent timestamps. The intuition is that (1) nearby people tend to participate in the same activity and (2) a person tends to perform the same activity in nearby time stamps. Such model can be formulated as follows.

Let x^i_t and y^i_t denote the data and collective activity labels associated with an individual person i at a certain time stamp t. Then the posterior probability over all activity labels y given all input x can be represented as:

$$P(y|x,l) \propto \prod_t \prod_i P(y^i_t|x^i_t) \prod_t \prod_{(i,j)\in E_s} \Phi_S(y^i_t, y^j_t; l^i_t, l^j_t) \prod_i \prod_t \Phi_T(y^i_{t-1}, y^i_t) \quad (9)$$

where l^i_t is the location of person i in t, E_s is the set of edges between people (Fig. 8), $P(y^i_t|x^i_t)$ is the unary probability estimate from Random Forest for a person i in time t, $\Phi_S(y^i_t, y^j_t; l^i_t, l^j_t)$ is the spatial pair-wise potential, and $\Phi_T(y^i_{t-1}, y^i_t)$ is the temporal pairwise potential. The temporal edges are established between temporally adjacent nodes of the same person. The two nodes in the same time-stamp are connected if they are close to each other (<2 m) in order to enforce similar labelling between the two. The maximum-a-posteri (MAP) solution of the MRF can be obtained by a Gibbs sampling procedure [5] given the parameters for the pairwise potentials. In [12], the temporal pairwise potentials are obtained by counting and normalize the co-occurrence of a pair of collective activity. The spatial potentials are estimated in a non-parametric way by collecting location difference oriented with respect to each person's pose for all activity pairs.

Table 1 Average classification results of various state-of-the-art [25] and baseline methods on the dataset [11] with 5 activities (left column) and 6 activities (right column). See text for details

Dataset	5 Activities (%)	6 Activities (%)
AC [25]	68.2	–
STL [11]	64.3	–
STL + MC [11]	65.9	–
STL + RF [12]	64.4	–
RSTV [12]	67.2	71.7
RSTV + MRF [12]	**70.9**	**82.0**

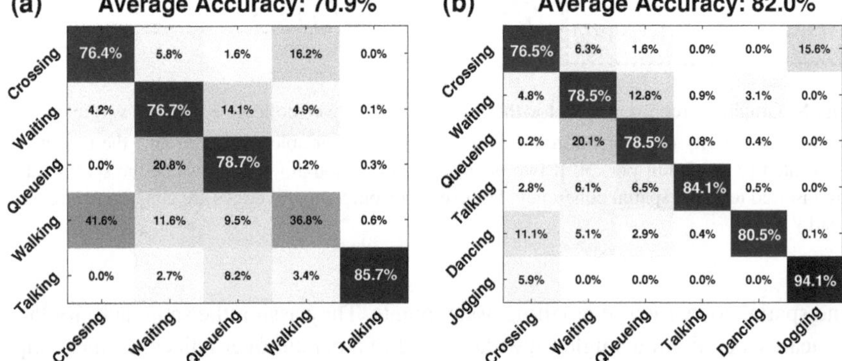

Fig. 9 The confusion tables using RSTV with MRF regularization on the dataset [11] with 5 activities (**a**) and 6 activities (**b**)

6 Experimental Results

In this section, we present an overview of the classification results obtained using the STL [11] and RSTV [12] crowd context descriptors. Both methods are evaluated using a publicly available dataset (Collective Activity Dataset [1]). Though there exist many different types of datasets for human activity recognition such as CAVIAR, IXMAS, KTH or UIUC, none of them are suitable for the proposed descriptors in that they focus on activities performed by a single or very few actors.

Dataset: There exist two versions of collective activity datasets [1]. The first version of the dataset is composed of five different collective activity categories, *Crossing, Standing, Queuing, Walking* and *Talking*. It includes 44 short video clips each of which is recorded from a real world scene with a variable number of people (see Fig. 3 for examples). The second version of the dataset includes 6 different collective activity categories, *Crossing, Standing, Queuing, Talking, Dancing* and *Jogging*. Similarly to the first version, the second version of the dataset has 74 short video clips with a variable number of people in the scene.

In both sequences, the videos were taken from a hand-held camera with a unpredictable camera motion incurred by jittering of the hand. Thus, the method of [10] is used in order to obtain the 3D trajectories of people in the videos. The anecdotal classification results obtained by [12] are shown in the Fig. 10.

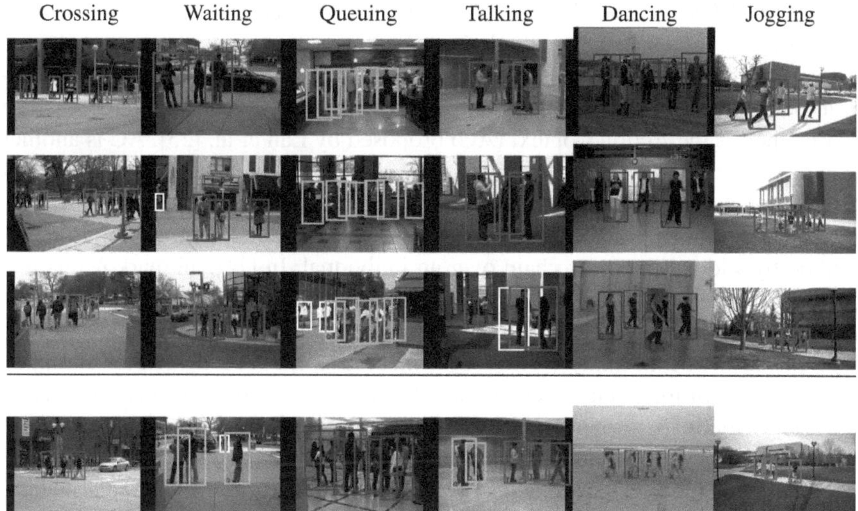

Fig. 10 Example results on the 6-category dataset [1] using RSTV with MRF. Top 3 rows show examples of good classification and bottom row shows examples of false classification. The labels X (*magenta*), S (*blue*), Q (*cyan*), T (*orange*), D (*red*), J (*green*) and NA (*white*) indicate *crossing*, *waiting*, *queuing*, *talking*, *dancing*, *jogging* and not assigned, respectively. When there is insufficient evidence to perform classification, the NA label is displayed. The misclassified results indicate that miss classifications mostly occur between classes with similar structure. This figure is best viewed in color

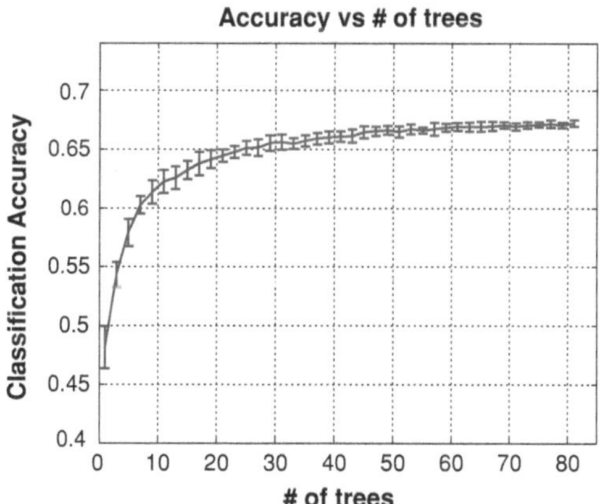

Fig. 11 Classification accuracy by RSTV using different number of trees. As the number of trees increases, the classification accuracy also improves and converges at around 60 trees. The 5-category dataset is used in this experiment. Vertical bars measure the standard deviation around the average classification accuracy

Activity Classification Results: Table 1 shows the comparison among several state-of-the-art and baseline methods for collective activity classification evaluated on the two collective activity datasets. Each row represents the overall classification accuracy and the columns represent different versions of the datasets. The first method is the Action Context (AC) proposed by Lan et al. [25]. AC is another type of contextual descriptor that accumulates the activity classifier confidence of both the anchor person and surrounding people. The second method is the STL descriptor equipped with SVM classifier. The third method augments the second method by adding a markov chain over an individual. In [11], an markov chain is introduced for each individual person in order to utilize temporal smoothness of the collective activity, i.e. a person doing *Crossing* tends to continue doing *Crossing* in next timestamps. The fourth method is the STL descriptor equipped with the random forest classifier. As noted in the comparison between the second and fourth methods, a mere replacement of the SVM classifier with a Random Forest does not yield an improvement in the collective activity classification. The fifth method is the RSTV and the last is RSTV equipped with MRF regularization. As shown in the table, the RSTV with MRF method gives the most robustness in collective activity classification thanks to the flexibility in learning the contextual information. All results presented are obtained using a leave-one-video-out training and testing scheme.

Figure 9 presents the confusion table for the collective activity dataset using the RSTV with MRF regularization. In the five category dataset experiment, the most confusion in classification occurs in discriminating the *Crossing* and *Walking* classes. This seems reasonable since the two classes share similar spatio-temporal properties. In the augmented 6 category experiment, the method produces more stable classification results since each collective activity category has distinctive spatio-temporal characteristics which can be more effectively captured by the crowd context descriptors.

Since each tree in RSTV forest is trained using the *bagging* procedure, each tree captures different spatio-temporal characteristics for each collective activity class. Thus having larger number of trees would provide more robust classification results in general. Such trend is shown in Fig. 11. When only few trees are used in the experiment, classification results by RSTV are rather unstable. As the number of trees increases, the classifier becomes more robust and converges to the best accuracy at 60 trees.

7 Conclusion

In this chapter, we have reviewed a recent formulation for classifying collective activities that takes advantage of the concept of *crowd context* and introduced two descriptors (STL [11] and RSTV [12]) to model the crowd context. Experimental evaluation indicates that the crowd context is a powerful and robust source of information to discriminate different types of collective activities.

References

1. Collective Activity Dataset. http://www.eecs.umich.edu/vision/activity-dataset.html
2. Amer, M.R., Todorovic, S.: A chains model for localizing participants of group activities in videos. In: Proceedings of International Conference on Computer Vision (ICCV) (2011)
3. Andriluka, M., Roth, S., Schiele, B.: People-tracking-by-detection and people-detection-by-tracking. In: CVPR (2008)
4. Belongie, S., Malik, J., Puzicha, J.: Shape matching and object recognition using shape contexts. PAMI **24**(4), 509–522 (2002)
5. Bishop, C.M: Pattern Recognition and Machine Learning. Springer, Berlin (2006)
6. Breiman, L., Cutler, A.: Random Forest. [online], marzec (2004)
7. Breitenstein, M.D., Reichlin, F., Leibe, B., Koller-Meier, E., Gool, L.V.: Robust tracking-by-detection using a detector confidence particle filter. In: ICCV (2009)
8. Chang, C.-C., Lin, C.-J.: LIBSVM: a library for support vector machines. Software available at http://www.csie.ntu.edu.tw/cjlin/libsvm (2001)
9. Choi, W., Pantofaru, C., Savarese, S.: Detecting and tracking people using an rgb-d camera via multiple detector fusion. In: Challenges and Opportunities in Robot Perception, ICCV, Nov 2011
10. Choi, W., Savarese, S.: Multiple target tracking in world coordinate with single, minimally calibrated camera. In: ECCV, Sept 2010
11. Choi, W., Shahid, K., Savarese, S.: What are they doing?: collective activity classification using spatio-temporal relationship among people. In: VSWS (2009)
12. Choi, W., Shahid, K., Savarese, S.: Learning context for collective activity recognition. In: CVPR (2011)
13. Dalal, N., Triggs, B.: Histograms of oriented gradients for human detection. In: CVPR (2005)
14. Dollar, P., Rabaud, V., Cottrell, G., Belongie, S.: Behavior recognition via sparse spatio-temporal features. In: VS-PETS (2005)
15. Ess, A., Leibe, B., Schindler, K., van Gool, L.: A mobile vision system for robust multi-person tracking. In: CVPR (2008)
16. Ess, A., Leibe, B., Schindler, K., van Gool, L.: Robust multi-person tracking from a mobile platform. PAMI **31**(10), 1831–1846 (2009)
17. Fanti, C., Zelnik-Manor, L., Perona, P.: Hybrid models for human motion recognition. In: CVPR, vol. 1, pp. 1166–1173, June 2005
18. Felzenszwalb, P., Girshick, R., McAllester, D., Ramanan, D.: Object detection with discriminatively trained part based models. PAMI **32**(9), 1627–1645 (2010)
19. Gorelick, L., Blank, M., Shechtman, E., Irani, M., Basri, R.: Actions as space-time shapes. Trans. Pattern Anal. Mach. Intell. **29**(12), 2247–2253 (2007)
20. Hakeem, A., Shah, M.: Learning, detection and representation of multi-agent events in videos. AI **171**, 586–605 (2007)
21. Intille, S., Bobick, A.: Recognizing planned, multiperson action. CVIU **81**, 414–445 (2001)
22. Khan, Z., Balch, T., Dellaert, F.: MCMC-based particle filtering for tracking a variable number of interacting targets. PAMI **27**, 1805–1819 (2005)
23. Kim, T., Wong, S.-f., Cipolla, R.: Tensor canonical correlation analysis for action classification. In: CVPR, June 2007
24. Kohli, P., Torr, P.H.S.: Dynamic graph cuts and their applications in computer vision In: Computer Vision: Detection, Recognition and Reconstruction, pp. 51–108 (2010)
25. Lan, T., Wang, Y., Mori, G., Robinovitch, S.: Retrieving actions in group contexts. In: International Workshop on Sign Gesture Activity (2010)
26. Lan, T., Wang, Y., Yang, W., Mori, G.: Beyond actions: discriminative models for contextual group activities. In: NIPS (2010)
27. Laptev, I., Lindeberg, T.: Space-time interest points. In: ICCV (2003)
28. Laptev, I., Marszalek, M., Schmid, C., Rozenfeld, B.: Learning realistic human actions from movies. In: CVPR (2008)

29. Leibe, B., Leonardis, A., Schiele, B.: Combined object categorization and segmentation with an implicit shape model. In: Statistical Learning in Computer Vision, ECCV (2004)
30. Li, R., Chellappa, R., Zhou, S.K.: Learning multi-modal densities on discriminative temporal interaction manifold for group activity recognition. In: CVPR (2009)
31. Liu, J., Ali, S., Shah, M.: Recognizing human actions using multiple features. In: CVPR (2008)
32. Liu, J., Kuipers, B., Savarese, S.: Recognizing human actions by attributes. In: Proceedings of the IEEE International Conference on Computer Vision and Pattern Recognition (2011)
33. Liu, J., Luo, J., Shah, M.: Recongizing realistic actions from videos "in the wild". In: CVPR (2009)
34. Liu, J., Shah, M., Kuipers, B., Savarese, S.: Cross-view action recognition via view knowledge transfer. In: Proceedings of the IEEE International Conference on Computer Vision and Pattern Recognition (2011)
35. Lu, W.-L., Little, J.J.: Simultaneous tracking and action recognition using the pca-hog descriptor. In: Proceedings of the 3rd Canadian Conference on Computer and Robot Vision (2006)
36. Lv, F., Nevatia, R.: Single view human action recognition using key pose matching and viterbi path searching. In: CVPR (2007)
37. Marszalek, M., Laptev, I., Schmid, C.: Actions in context. In: CVPR (2009)
38. Niebles, J.C., Chen, C.-W., Fei-Fei, L.: Modeling temporal structure of decomposable motion segments for activity classification. In: Proceedings of the 12th European Conference of Computer Vision (ECCV), Crete, Greece, Sept 2010
39. Niebles, J.C., Wang, H., Fei-Fei, L.: Unsupervised learning of human action categories using spatial-temporal words. IJCV **79**, 299–318 (2008)
40. Patron, A., Marszalek, M., Zisserman, A., Reid, I.: High five: Recognising human interactions in tv shows. In: Proceedings of the BMVC (2010)
41. Pirsiavash, H., Ramanan, D., Fowlkes, C.: Globally-optimal greedy algorithms for tracking a variable number of objects. In: CVPR (2011)
42. Ramin Mehran, A.O., Shah, M.: Abnormal crowd behavior detection using social force model. In: CVPR (2009)
43. Ryoo, M.S., Aggarwal, J.K.: Spatio-temporal relationship match: Video structure comparison for recognition of complex human activities. In: ICCV (2009)
44. Ryoo, M.S., Aggarwal, J.K.: Stochastic representation and recognition of high-level group activities. IJCV **93**(2), 183–200 (2010)
45. Savarese, S., DelPozo, A., Niebles, J., Fei-Fei, L.: Spatial-temporal correlatons for unsupervised action classification. In: WMVC (2008)
46. Song, Y., Goncalves, L., Perona, P.: Unsupervised learning of human motion. PAMI **25**(25), 1–14 (2003)
47. Swears, E., Hoogs, A.: Learning and recognizing complex multi-agent activities with applications to american football plays. In: WACV (2011)
48. Turaga, P., Chellappa, R., Subrahmanian, V.S., Udrea, O.: Machine recognition of human activities: a survey. IEEE Trans. Circuits Syst. Video Technol. **18**(11), 1473–1488 (2008)
49. Viola, P., Jones, M., Snow, D.: Detecting pedestrians using patterns of motion and appearance. In: ICCV (2003)
50. Wojek, C., Walk, S., Roth, S., Schiele, B.: Monocular 3d scene understanding with explicit occlusion reasoning. In: CVPR (2011)
51. Wojek, C., Walk, S., Schiele, B.: Multi-cue onboard pedestrian detection. In: CVPR (2009)
52. Wong, S.-F., Kim, T.-K., Cipolla, R.: Learning motion categories using both semantic and structural information. In: Proceedings of IEEE Conference on Computer Vision and Pattern Recognition (CVPR) (2007)
53. Wu, B., Nevatia, R.: Detection and tracking of multiple, partially occluded humans by bayesian combination of edgelet based part detectors. IJCV **75**(2), 247 (2007)
54. Yao, A., Gall, J., Van Gool, L.: A hough transform-based voting framework for action recognition. In: CVPR (2010)
55. Yu, T., Kim, T.-K., Cipolla, R.: Real-time action recognition by spatiotemporal semantic and structural forest. In: Proceedings of British Machine Vision Conference (BMVC) (2010)

56. Zhang, L., Li, Y., Nevatia, R.: Global data association for multi-object tracking using network flows. In: CVPR (2008)
57. Zhou, B., Wang, X., Tang, X.: Understanding collective crowd behaviors: learning mixture model of dynamic pedestrian-agents. In: CVPR (2012)

Event Based Switched Dynamic Bayesian Networks for Autonomous Cognitive Crowd Monitoring

Simone Chiappino, Lucio Marcenaro, Pietro Morerio and Carlo Regazzoni

Abstract Human behavior analysis is one of the most important applications in Intelligent Video Surveillance (IVS) field. In most recent systems addressed by research, automatic support to the human decisions based on object detection, tracking and situation assessment tools is integrated as a part of a complete cognitive artificial process including security maintenance procedures actions that are in the scope of the system. In such cases an IVS needs to represent complex situations that describe alternative possible real time interactions between the dynamic observed situation and operators' actions. To obtain such knowledge, particular types of Event based Dynamic Bayesian Networks E-DBNs are here proposed that can switch among alternative Bayesian filtering and control lower level modules to capture adaptive reactions of human operators. It is shown that after the off line learning phase Switched E-DBNs can be used to represent and anticipate possible operators' actions within the IVS. In this sense acquired knowledge can be used for either fully autonomous security preserving systems or for training of new operators. Results are shown by considering a crowd monitoring application in a critical infrastructure. A system is presented where a Cognitive Node (CN) embedding in a structured way Switched E-DBN knowledge can interact with an active visual simulator of crowd situations. It is also shown that outputs from such a simulator can be easily compared with video signals coming from real cameras and processed by typical Bayesian tracking methods.

Keywords Crowding · Dynamic Bayesian Networks · Cognitive Systems · Bio-inspired learning

S. Chiappino · L. Marcenaro (✉) · P. Morerio · C. Regazzoni
Signal Processing and Telecommunications Group, Department of Naval, Electrical, Electronic and Telecommunication Engineering, University of Genoa, Via All'Opera Pia 11A, 16145 Genoa, Italy
e-mail: mlucio@dibe.unige.it

Augment Vis Real (2014) 6: 93–122
DOI: 10.1007/8612_2012_8

Published Online: 14 September 2012

1 Introduction

A lot of works have been devoted in the last decade to link traditional computer vision tasks to high-level context aware functionalities such as scene understanding, behavior analysis, interaction classification or recognition of possible threats or dangerous situations [1–4].

Among the several disciplines which are involved in the design of next generation security and safety systems, cognitive sciences [5] represent one of the most promising in terms of capability of provoking improvements with respect to state of the art. As a matter of fact, several recent studies have proposed the application of smart functionalities to camera and sensor networks in order to move from object recognition paradigm to event/situation recognition one [6]. The application of bio-inspired models to safety and security tasks represents a relevant added value. In fact, the capability not only of detecting the presence of an intruder in a forbidden area or recognizing the trajectory of an object in an urban scenario (e.g. a baggage in a station or a car on the road) but also of interpreting the behavior of the entity in the monitored scene or properly selecting events of interest (up to anomalous events) with respect to normal situations. In addition, to efficiently exploit cognitive capabilities in an intelligent sensor network, the role of data fusion algorithms is crucial [7, 8].

In the literature, several works deal with data fusion problem applied to heterogeneous sensors both for security [9, 10] and safety tasks [11, 12]. In this work, the features of a cognitive-based framework, inspired by the previously cited concepts, are described and the application of the proposed architecture to crowd analysis is presented.

The proposed Cognitive Node (CN) can be applied to the crowd analysis domain to effectively join technical and social aspects related to the behavior of groups of people. In this scenario the goal of the system is to analyze and classify crowd interactions in order to maintain a proper security level in the monitored area and to put in action effective countermeasures in case of detection of panic or overcrowding situations. Simulated data is employed for testing.

2 Crowd: Modeling, Simulation and Monitoring

Crowd Monitoring, Simulation and Modeling are strongly related fields and this is the reason why they are often discussed altogether. First of all, the mere need for simulating and monitoring crowd raises the issue of modeling its behavior: crowds obviously need to be given a dynamic evolution model to be simulated; also, a dynamic model is often needed to improve Crowd monitoring application performances through Bayesian filtering; then again, simulations are often necessary in order to test Crowd Monitoring algorithms; eventually, Crowd Monitoring can provide valuable hints on how to effectively model and describe crowds.

A comprehensive traction of such interconnected fields is given in the following by trying to stress links, similarities, differences and synergies between them.

2.1 Scale Issues

Sure enough, one should ask himself what a crowd is, before starting discussing about it. The way people define a crowd obviously depends on the area in which the crowd itself is investigated, and thus many different definitions can be found in literature. However, any definition one could try to give can hardly avoid describing crowd in terms of its components, namely the people which it is formed by. This remark may sound trivial, but it has deep implications in the way a crowd is depicted. In particular, it raises the issue of choosing between a local description of it and a global one. A local description of a crowd relies on the features associated to each member, such as positions, speeds, directions, motivations, destinations etc. A global (holistic) description, on the other hand, relies on features that can be associated to the crowd as a single entity, such average density, the entropy, the average shift in some direction, the displacement etc. Global features can in general be derived from local ones, by averaging or integrating local quantities. The opposite, on the contrary, never happens. However, it is not only a matter of scale at which the crowd is analyzed, but rather of the additional amount of information stored in local quantities compared to global ones.

A nice parallel example comes from the well-known thermodynamics, where global quantities, such as energy, pressure and temperature of gases can in principles be derived from the average kinetic energy of its molecules: by knowing the exact behavior of each single molecule in the gas one can derive the temperature, while the opposite calculation is not possible, as information is lost by averaging over all molecules.

However, in both the cases of crowd and thermodynamics, it is not always possible to access local information entirely, while global quantities can be easily gathered. For example, in a video surveillance framework, it is unrealistic to track every single person in a high density crowded scene, especially if a single camera is available: the visual information gathered by the camera sensor is simply not enough to accomplish such a task. This kind of considerations has led to suggest approaches such as the one proposed in [29], in which a very subtle analysis is performed, taking into account a global macroscopic scale, a middle mesoscopic scale and eventually a local microscopic scale in a hydrodynamics-inspired framework (here again physics is of great help).

A perfectly specular approach is on the contrary often adopted in simulating and also modeling crowds. Here an underlying model can be designed in order to model the fine-scale behavior of each crowd member, in order to reproduce (simulate) some desired macroscopic behavior. This approach can on the one hand be really helpful in fine tuning macroscopic simulation outputs by correcting microscopic local parameters in the model. On the other hand it can be a very effective way to validate the accuracy of models, as it gives a way to check their accuracies in reproducing global crowd behaviors.

2.2 Crowd Monitoring

The crowd phenomenon has recently increasingly attracted the attention of worldwide researchers in video surveillance and video analysis [13] and nowadays an extremely prolific literature is growing on the subject. Different implications related to crowd behavior analysis can be considered, since both technical and social aspect is still under researchers' investigation.

On the one hand, researchers focusing on psychology and sociology domains consider crowd behavior modeling as a social phenomenon. Several examples can be found in the open literature dealing with the role and the relevance of human interaction factors in characterizing the behavior of a crowd. In [14], a simulation-based approach to the creation of a population of pedestrians is proposed. The authors aim at modeling the behavior of up to 10,000 pedestrians in order to analyze several movement patterns and people reactions typical of an urban environment. The impact of emotions of individual agents in a crowded area has been investigated also by Liu et al. [15] in order to simulate and model the behavior of groups of people. As well, Handford and Rogers [16] have recently proposed a framework for modeling drivers' behavior during an evacuation in a post disaster scenario taking into account several social factors which can affect their behavior in following a path to reach a safe spot.

To the other hand, technical aspects in crowd behavior analysis applications mainly focus on the detection of events or the extraction of particular features exploiting computer vision based algorithms. An estimation of the number of people in a crowd can be performed by computing the number of foreground and edge pixels. Davies et al. propose a system using Fourier transform for estimating the motion of the crowd [17]. Many researchers tried to use segmentation and shape recognition techniques for detecting and tracking individuals and thus estimating the crowd. However this kind of approach can hardly be applied to overcrowding situations where people are typically severely occluded [18, 19]. Neural networks are used in [20] for estimating crowd density from texture analysis, but in this case an extensive training phase is needed for getting good performances. A Bayesian model based segmentation algorithm was proposed in [21]; this method uses shape models for segmenting individual in the scene and is thus able to estimate the number of people in the crowd. The algorithm is based on Markov chain Monte Carlo sampling and it is extremely slow for large crowds. Optical flow based technique is used in [22, 23], while Rahmalan et al. [24] proposed a computer vision-based approach relying on three different methods to estimate crowd density for outdoor surveillance applications.

As a matter of fact, the combination of technical and social aspects can represent an added value with respect to the already presented works. A first example can be found in [25] where authors exploit a joint visual tracking-Bayesian reasoning approach to understand people and crowd behavior in a metro station scenario. More recently [26–29], a social force model describing the interactions among the individual members of a group of people has been proposed to detect abnormal events in crowd videos. Here people are treated as interacting particles subject to internal and external physical forces which determine their motion. At the same time social and psychological aspects are taken into account in modeling such indeed "social" forces, showing the effectiveness of a synergic multidisciplinary approach to the problem.

2.3 Simulating Crowds

Graphical or symbolical simulation of moving crowds is a continuously evolving field which involves research groups all around the world in many different areas, such as entertainment industry (videogames and motion-picture), police and military force training (manifestations and riots simulations), architecture (buildings and cities design), traffic control (crossovers and walking paths), security sciences (evacuation of crowded environments) and sociology (behavior studies). Simulation of crowds meets the needs for crowd observation data that are often hard or even impossible to gather directly and is also often necessary in the design stage of security and surveillance systems.

Here again different application areas obviously show different approaches to the problem. Basically, these approaches can be divided into two main categories. The first is mostly focused on behavioral aspects of the crowd, while neglects visual output quality. Crowd members can be schematically represented as dots or stylized shapes or even melt together in a rougher framework, wherever only a global point of view is needed. Here, only realism of dynamics is stressed. The second approach, on the contrary, is centered on visual effects and it is not really concerned with an appropriate modeling of the real behavior. A well balanced integration of realism in the behavior of the crowd and in the visualization of it is also often needed, at least to some extent, as in the case here presented. This will be discusses in details in the following.

As mentioned at the beginning of this section, crowds need to be given an underlying dynamical model in order to be simulated. Actually, such a model is inherently in charge of depicting the evolution of some crowd features only. This raises again the issue of how to describe crowds. This includes a selection of the features one is interested in simulating, but also of the scale at which the model has to lie, in order to effectively describe the formers. Namely, a microscopic model could be given the task of simulating features at a more global level, while the opposite way is hardly practicable.

3 The Cognitive Model

The proposed approach to IVS has been implemented according to a bio-inspired model of human reasoning and consciousness grounded on the work of the neuro-physiologist Damasio [5]. Damasio's theories describe the cognitive entities as complex systems capable of incremental learning based on the experience of the relationships between themselves and the external world. Two specific brain devices can be defined to formalize the above concept called proto-self and core-self. Such devices are specifically devoted to monitor and manage respectively the internal status of an entity (proto-self) and the relationships with the external world (core-self). Thus, a crucial aspect in modeling a cognitive entity following Damasio's model is first of all represented by the capability of accessing entity's internal status and secondly by the knowledge and analysis of the surrounding environment. This approach can be mapped into a sensing framework by dividing the sensors into endo-sensors (or proto-sensors) and eso-sensors (or core-sensors) as they monitor, respectively, the internal or external state of the interacting entities. The core of the proposed architecture is the so called CN. It can be considered as a module that is able to receive data from sensors, to process them for finding potentially dangerous or anomalous events and situations, and, in some cases, to interact with the environment itself or contact the human operator.

3.1 Cognitive Cycle for Single and Multiple Entities Representation

Within the proposed scheme the representation of each entity has to be structured into a multi-level hierarchical way. As a whole, the closed processing loop realized by the CN in case of a given interaction between an observed object and the system can be represented by means of the so-called cognitive cycle (CC—see Fig. 1) which is composed of four main steps:

- Sensing: the system has to continuously acquire knowledge about the interacting objects and about its own internal status.
- Analysis: the collected raw knowledge is processed in order to obtain a precise and concise representation of the occurring causal interactions.
- Decision: the precise information provided by analysis phase is processed and a decision strategy is selected according to the goal of the system.
- Action: the system put into practice the configuration provided by the decision phase under the form of a direct action over the environment or of a message provided to the user.

In addition, the learning phase is continuous and involves all the stages (within certain limits) of the CC. Thus, the CC can be viewed as a dispositional embodied description of an object as it includes reactions it generates in the cognitive system,

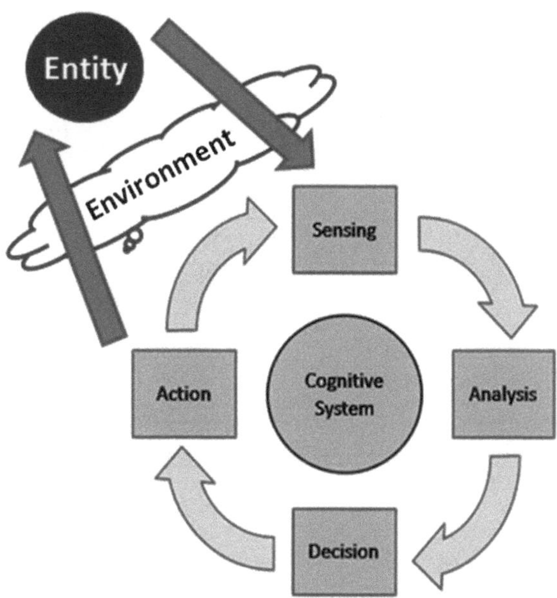

Fig. 1 Cognitive cycle (single object representation)

i.e. to possible actions that the system can plan and perform when a situation involving that object is observed and predicted. According to this statement, it is possible to refer to the representation model depicted in Fig. 1 as to an Embodied Cognitive Cycle (ECC). With respect to security and safety domains, in which the ECC is here applied, the above mentioned embodied description is associated to a precise objective: to maintain stability of the equilibrium between the object and the environment (i.e. maintenance of the proper level of security and/or safety). As a consequence, each entity is provided by a 'security/safety oriented ECC (S/S-ECC)' which is representative of the entity itself within the CN. The mapping of the S/S-ECC onto the cognitive node chain shown in Fig. 2 can be viewed as the result of the interaction between two entities, each one described as a cognitive cycle too. In particular, if the external object (eso) and the internal autonomous system (endo) are represented as a couple of Interacting Virtual Cognitive Cycles (IVCC). The IVCCs can be matched with the CN structure (i.e. the bottom-up and the top-down chains) by associating parts of the knowledge related with the different ECC phases to the multilevel structure processing parts of the CN (Fig. 3).

More in detail, the representation model of the ECC (top left corner of Fig. 3) is centered on the cognitive system that can be considered by itself as a cognitive entity. Therefore, it is possible to map the proposed representation as in the top right corner of Fig. 3, where two IVCCs, the one representing the entity (or object—IVCCo) and the other representing the cognitive system (IVCCs), interact in a given environment. In this model, the sensing and action blocks of the IVCCs correspond to the sensing and action blocks of the ECC (see bottom right corner of the figure). However, in the IVCCs, such blocks assume a parallel virtual

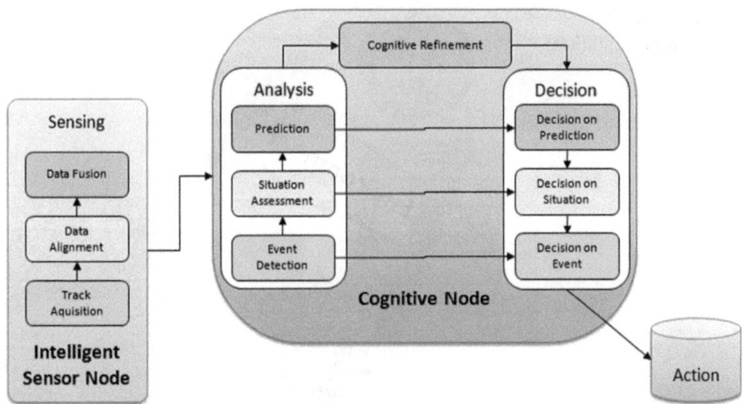

Fig. 2 Cognitive node: bottom-up analysis and top-down decision chain

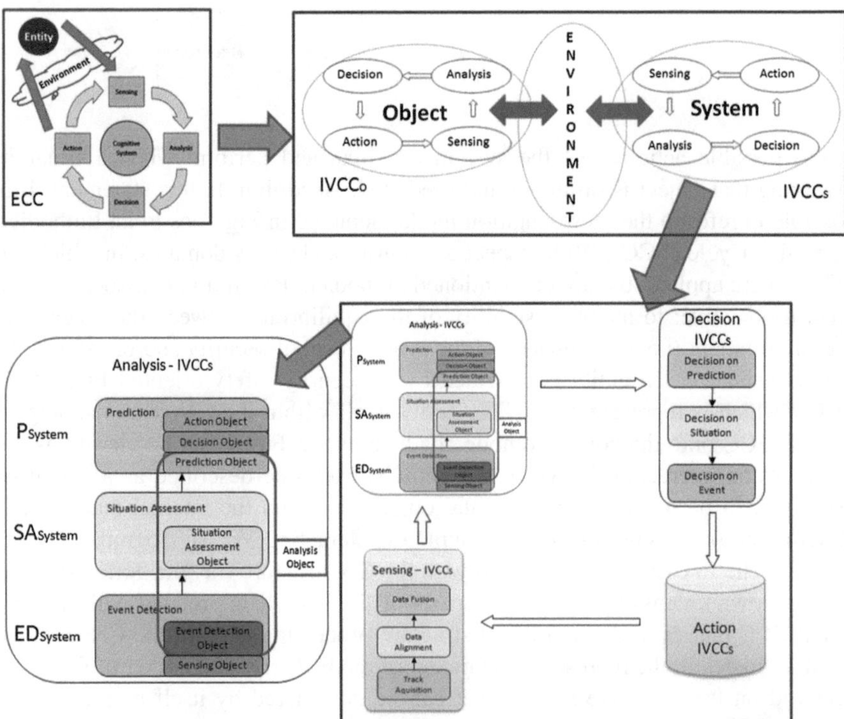

Fig. 3 Embodied cognitive cycle, interactive virtual cognitive cycles and cognitive node matching representation

representation of the physical sensing and action observed corresponding respectively to the Intelligent Sensing Node and the Actuator blocks in the general framework.

The proposed interpretation of the matching among the embodied cognitive model, the interactive virtual cycles representing the entities acting in the environment (including the system) and the CN allows considering the CN as a universal machine for processing ECCs with respect to a large variety of application domains. In general, each ECC starts with Intelligent Sensor Node (ISN) data including an interacting entity (eso-sensor) and a system reflexive observation (endo-sensor). The observed data (acquired under system viewpoint) are considered in two different perspectives (the object and the system) by creating a description of the current state of the entities using knowledge learned in previous experiences. Such process happens at event detection and situation assessment sub-blocks. Then, a prediction of future actions taken by the IVCCo, contextualized with the self-prediction of future planned actions of the system, occur at prediction sub-block. The use of the knowledge of the IVCCo ends at this stage. Finally, the IVCCs is completed by adjusting plans of the system in the representation of its decision and action phases that are, as stated above, a parallel virtualization of the ECC.

In addition, it is relevant to briefly point out that a similar decomposition can be adopted in the case when two interactive entities are observed. The description of the interacting subjects can be modeled observing that the two entities can form a single meta-entity to which is associated a meta-cognitive cycle interacting with the autonomous system. As the meta-entity (ME) can simply be considered as a composition of the two cognitive cycles associated to the initial entity couple.

The advantage of the proposed representation, involving the description of an Embodied Cognitive Cycle by means of an IVCC couple is that the same mechanism used to represent the interaction of a ME with the autonomous system can be also used to represent the interaction between two observed entities forming an observed meta-entity.

3.2 The Cognitive Node

The general architecture of the CN, which was briefly introduced at the beginning of Sect. 3, is depicted in Fig. 4.

Intelligent sensors are able to acquire raw data from physical sensors and to generate feature vectors corresponding to the entities to be observed by the CN. Acquired feature vectors must be fused spatially and temporally in the first stages of the node, if they are coming from different sources.

As already mentioned, the CN is internally subdivided into two main parts: the analysis and the decision blocks linked through the cognitive refinement block. Analysis blocks are responsible for organizing sensors data and finding interesting or notable configurations of the observed entities at different levels. Those levels can communicate directly with the human operator through network interfaces in the upper part of Fig. 4. This is basically what can be done by a standard signal processing system being able to alert a supervisor whenever a specific event is

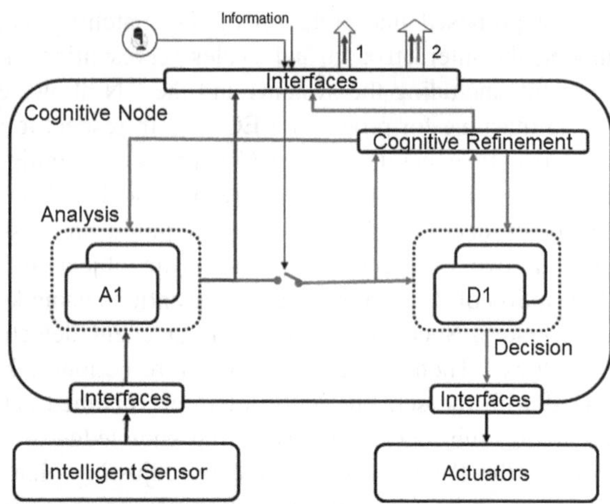

Fig. 4 Cognitive node architecture

detected. A prediction module is able to use the stored experience of the node through the internal autobiographical memory for estimating a possible evolution of the observed environment. All the processed data and predictions generated by the analysis steps are used as input of the cognitive refinement block. This module can be seen as a surrogate of the human operator: during the configuration of the system it is able to learn the best way to interact with the environment. In the on-line phase, the acquired knowledge migrates to the decision block, which is responsible for autonomously choosing the best available strategy to avoid dangerous situations. All the decisions taken by the CN are made according to the stored knowledge, with the intent of maintaining the environment in a controllable, alarm-free state (for we suppose the system has been trained to do so). A human operator always has the possibility of deciding and completely bypassing the automatic system or to be forced to acknowledge each single action that the CN is transmitting to the guarded environment.

4 Information Extraction for Probabilistic Interaction Model

Interactions between two entities can be described in terms of mathematical relationships. However, such a mathematical description must rest on a feature extraction phase, which is addressed to get relevant information about the entities.

This section is devoted to the analysis of the main features that allow to design of a probabilistic model able to learn interactions.

After information is extracted, Dynamic Bayesian Networks (DBNs) [30] can be used to represent cognitive cycles and IVCCs based on an algorithm, called autobiographical memory [31], and provide a tool for describing embodied objects within the CN in a way that can allow incremental learning from experience. It could be noted that also interactions between the operator and the system can be represented as an IVCC. In that case, the operator-system interaction can be differently used as an internal reference for the CN as the operator can be seen as a teaching entity addressing most effective actions towards the goal of maintaining security/safety levels during the learning phase. This learning phase represents an effective knowledge transfer from human operator towards an automatic system. A proposed framework for information extraction is composed of two main blocks: Data Fusion (DF) and Event Detection (ED). DF involves the following phases: source separation, feature extraction and dimensional reduction. The ED block extracts information related to changings in the signals acquired by sensors. Events will be eventually defined, in order to develop some specific probabilistic models.

4.1 Data Fusion

The data fusion module is able to receive data from intelligent sensors on the field, and to fuse them from a temporal and spatial point of view. Consider a set of S intelligent sensors: each $k \in S$ sends to the cognitive node a vector of features $\vec{x}(k,t) = \{x_1, x_2, \ldots, x_{N_k}\}$ where $k = \{1, 2, \ldots, S\}$ at time instant t. Intelligent sensors send feature vectors asynchronously to the CN, that must be able to register them temporally and spatially before sending data to upper level processing modules.

From a temporal point of view, the DF module collects and stores into an internal buffer all newest measurements $\vec{x}_{k,t_k^*}$ from the kth intelligent sensor at a time instant t_k^*. Data acquisition time can vary from sensor to sensor.

As soon as a new feature vector is acquired from sensor k, the data fusion module can compute an extended feature vector by combining all measurements from all considered intelligent sensors $\hat{\varphi}(\hat{t}) = f\left(\vec{x}_{1,t_1^*}, \vec{x}_{2,t_2^*}, \ldots, \vec{x}_{S,t_S^*}\right)$, where $\hat{t} \geq \{t_1^*, t_2^*, .., t_S^*\}$.

The analytic expression of the fusion function $\varphi(\hat{t})$, depends on the physical relationship between measured quantities and cannot be studied with a generic approach. In the following scenario, feature vectors are mainly generated by (simulated) video analytics algorithms that are able to process images acquired from video–surveillance cameras and extract scene descriptors (e.g. trajectories of moving objects, crowd densities, human activity related features, etc.). The fusion algorithm must be designed for being able to combine together all the sensor data from the guarded environment. If a set of disjoint video sensors is considered, the DF algorithm will find the union of considered feature vectors, thus giving to the upper modules of the cognitive node a more complete description of the considered world. The output feature vector can in general be written as:

$$\vec{x}(t) = \{\vec{x}_C, \vec{x}_P\} = \{\vec{x}_{C_1}, \vec{x}_{C_2}, \ldots, \vec{x}_{C_n}, \vec{x}_{P_1}, \vec{x}_{P_2}, \ldots, \vec{x}_{P_m}\} \tag{1}$$

In Eq. (1), $\vec{x}_C$ identifies features related to so-called *core* objects, i.e., entities that are detected within the considered environment but that are not part of the internal state of the system itself. Vector $\vec{x}_P$ identifies *proto* object features that are specific for entities that can be considered as part of the internal state of the CN.

4.2 Event Detection

The event detection step can be divided into an off-line and an on-line phase. During the learning off-line stage, temporally and spatially aligned feature vectors that are received from the data fusion module, are used to train an unsupervised classifier, a Self Organizing Map (SOM) [32] in this case, which is employed to convert the multidimensional proto and core feature vectors $\vec{x}_P(t)$ and $\vec{x}_C(t)$ to a lower M-dimensional map layer. By choosing $M = 2$, the SOM is referred to as a 2D SOM. The input vectors are clustered according to their similarities and to each cluster is assigned a label. Similarity is measured by means of some arbitrary distance metric. The choice of SOMs to perform feature reduction and clustering processes is due to their capabilities to reproduce in a plausible mathematical way the global behaviour of the winner-takes-all and lateral inhibition mechanism shown by distributed bio-inspired decision mechanisms.

The clustering process, applied to internal and external data allows one to obtain a mapping of proto and core vectors $\vec{x}_P(t)$ and $\vec{x}_C(t)$ in 2D vectors, corresponding to the positions of the neurons in the SOM map, that we call, respectively, proto *Super-states* $S_{i,P}$ and core *Super-states* $S_{i,C}$. Each cluster of Super-states, deriving from the SOM classifiers, is then associated with a label:

$$S_{i,P} \rightarrow \sigma_i^P, \;\; i = 1, \ldots, N_P; \quad S_{i,C} \rightarrow \sigma_i^C, \;\; i = 1, \ldots, N_C \tag{2}$$

where N_P and N_C are the maximum number of the proto and core Super-states labels, respectively. The dimension of the two label spaces actually correspond to the area of the 2D SOMs.

Then, by sequentially analysing the dynamic evolution of Super-states, proto and core events can be detected. The resulting information becomes an approximation of what Damasio calls the *Autobiographical Memory* where the interaction between user and system is memorized. The output of the off-line process is a list of labels corresponding to clusters within the reduced features space. The module also considers dynamic aspects of the evolution of clustered features: transition probabilities between different clusters are computed from learned data, so that the outcome of the training process can be ideally compared to the construction of a probabilistic model.

In the next section a proposed model based on Dynamic Bayesian Network is shown. This structure memorizes sequences of proto and core events to describe interactions. Events (i.e. state changes) are here considered instead of simply states, since they can be located in time and can then better describe cause-effect relationships. During the on-line phase, input feature vectors from DF block (1), are processed and a set of events is generated.

5 Bayesian Model for Interaction

In this section, a new type of Coupled Event based DBNs (C E-DBNs) is presented, in order to provide an efficient knowledge representation for modeling interactions between cognitive entities.

The basic idea is to introduce a different representation of proto and core events and their temporal evolution with respect to the one used in [33]. Within each single E-DBN, variables that describe events and the time at which they occur are explicitly represented and positioned according to a modified logical ordering. In a classic DBN, each slice is generally used to represent a set of variables associated with a given time instant; in this case, nodes represent events at a given time instant (slice), while links inter-slices represent dynamic dependencies between events at consecutive, time instants. This model is useful when a one-time step Markovianity can be hypothesized among events. However, this assumption often does not hold: event variations can be represented as sparse time series and one can generally observe the sequence of events to fix probabilistic dependencies among successive events without being able to deterministically fix the temporal range at which they occur. Having fixed a couple of successive events, time instants at which they occur can be considered as random variables and a temporal window can be found beyond which causality can be assumed to disappear and independence holds. If such a model is available, where Markovianity order is extended to the size of the temporal window, more structured predictions can be obtained at the expenses of a higher computational load. However, sparsity of event time series makes reduces computational load. In the following a model is proposed that is embedded in a new type of Coupled E-DBN, defined as Run Length Coupled Event DBN (RLCE-DBN).

5.1 Event Stream Modeling

Let us assume the reduced dimensionality state label obtained by quantizing core/proto observation time series onto a reduced dimensionality vectorial series as belonging to the set $\bar{S}_j = \{S_{1,j}, \ldots, S_{N_j j,}\}$. $\bar{S}_j$ is defined as the set of possible *Super state labels* [34, 35] observed at each time by a system where each $S_{i,j}$ denotes the ith component associated either with core source (if $j = C$) or with proto source (if $j = P$).

Each Super state observed at *m*th time t_m along the reduced dimensionality vector time series can be expressed either as a function $S_{i,j}(t_m)$ or as a associative pair (σ^j_m, t_m) where $\sigma^j_m \in \bar{S}_j$ is a semantic Super State Label observed at time t_m and one can write: $S_{i,j}(t_m) \mapsto (\sigma^j_m, t_m)$. In this way, a couple of linked variables are needed to express the super state with the advantage of explicitly representing time.

Let us define a label super state series as a temporal series of state variables $\Sigma_j = \{(\sigma^j_m, t_m)\colon m = 0...M-1, \sigma^j_m \in \bar{L}_j, t_0 = T_0, t_m = T_0 + m\Delta t\}$; where Δt is a uniform sampling step.

Following the above notation, an event A^j_m describing a Super State change from σ^j_{m-1} at time t_{m-1} to σ^j_m at t_m can be described as: $A^j_m = (\sigma^j_m, \sigma^j_{m-1}, t_m)$ where $\sigma^j_m, \sigma^j_{m-1} \in \bar{S}_j$. Let us now consider two alphabets from which state transitions estimated by Y can take values, $\mathcal{E}_j = \{\mathrm{a}^{j,1}, \ldots, \mathrm{a}^{j,V_j}\}$, where $j = P, C$, $\mathrm{a}^{j,i}$ is an event-symbol describing a specific label super state transition and V_j is the number of events. By using $\mathcal{E}_j$ it is possible to define the event in a synthetic way as: $A^j_m = (\mathrm{a}^j_m, t_m) : \mathrm{a}^j_m \in \mathcal{E}_j$. In general the ith a^j_m symbol occurs at time $t_m = t_{m-1} + \Delta t = m\Delta t$. In general, a^j_m can represent both a label change or in persistence in the same super state. In this latter case, we will write $\mathrm{a}^j_m = \emptyset$.

From a given state series Σ_j, an event sequence can be defined as $\Psi_j = \{A^j_m, m = 1...M-1\}$, by applying a simple differential production rule on a couple of consecutive time instants. This is equivalent to apply a mobile window $W_m(.,.)$ operator, called *Homogeneous Causality Window,* into two homogeneous (e.g. proto–proto or core–core events) consecutive random states of Σ_j series; therefore, $A^j_m = W_m((\sigma^j_m, t_m), (\sigma^j_{m-1}, t_{m-1}))$ allows one to obtain:

$$A^j_m = \begin{cases} (\alpha^j_m, t_m) : \mathrm{a}^j_m = \alpha^j_m;\text{ with } \alpha^j_m = (\sigma^j_m, \sigma^j_{m-1}) \text{ if } \sigma^j_m \neq \sigma^j_{m-1} \\ (\emptyset^j_m, t_m) : \mathrm{a}^j_m = \emptyset^j_m \text{ with } \emptyset^j_m = (\sigma^j_m, \sigma^j_{m-1}) \text{ if } \sigma^j_m = \sigma^j_{m-1} \end{cases} \tag{3}$$

where $\emptyset^j_m$ is a not relevant event, while α^j_m defines a generic relevant proto/core event. It should be noted that if a window of the same extension is applied to two random variables in two adjacent slices of a state based DBN it can be used to express a first-order Markovianity dependency.

5.2 Event Based Run Length Encoding

If one observes that often event changes of the reduced dimensionality vector are sparse in time as they occur rarely with respect to the uniform sampling time step, it can be useful to define a run length encoding RLE function $\Upsilon_j = R(\Psi_j)$ of event sequence Ψ_j as follows.

$$\Upsilon_j = \left\{x^j_1 \mathrm{a}^j_{m_1} ! x^j_2 \mathrm{a}^j_{m_2} ! \ldots\ldots ! x^j_N \mathrm{a}^j_{m_N}\right\} \tag{4}$$

The mark “!” is a redundant separator between a RLE element $\left(x^j_k \mathrm{a}^j_{m_k}\right)$ and a successive null event sequence. In Υ_j each element is described by $\left(x^j_k \mathrm{a}^j_{m_k}\right)$, where

x_k^j is the number of successive occurrences of an event $a_{m_k}^j$. This number is a random variable, which can be used to model probabilistic time lags between events, representing causes and their effects in a generic time series. It can be also noted that if the initial time t_0 of RLE event sequence is known, the absolute time index in Σ_j and Ψ_j can be recovered as $t_{m_k} = t_0 + \sum_{n=1}^{m_k} x_n^j \Delta t$.

Due to definition it holds that $a_{m_k}^j \neq a_{m_{k+1}}^j, \forall k$. It is also clear that it must be $x_k^j = 1$ if $a_{m_k}^j \neq \emptyset$. Otherwise, if $a_{m_k}^j = \emptyset$ it should hold that $x_k^j \geq 1$.

In this latter case, this means that in the event stream Ψ_j there will be no super state changes for x_{k-1}^j times before a non-null event $a_{m_k}^j$ is observed at time t_{m_k}. For example, as a consequence, it can be written that if $a_{m_2}^j = \emptyset$ then a variable $r_3^j = x_1^j + x_2^j + x_3^j = x_2^j + 2$ can be defined to describe observed time delay among $a_{m_3}^j$ and $a_{m_1}^j$, such that: $r_3^j \Delta t = t_{m_3} - t_1 = t_0 + \sum_{n=1}^{m_3} x_n^j \Delta t - \left(t_0 + \sum_{n=1}^{m_1} x_n^j \Delta t\right) = \left(x_3^j + x_2^j\right)\Delta t$. If $t_0 = 0$ and a discrete time series is fixed where $t_{m_k} = m_k$ indicates the position of each element in the series. More in general if $a_{m_{k-1}}^j = \emptyset$ one can write:

$$r_k^j \Delta t = \sum_{n=m_{k-1}}^{m_k} x_n^j \Delta t = t_{m_k} - t_{m_{k-2}} = (m_k - m_{k-2})\Delta t \tag{5}$$

Using RLE sequences allows one to highlight a different kind of first-order Markovianity dependencies between not null events: only relevant events that occur in consecutive positions of the RLE coded sequence, eventually separated by a null event, can be considered do be directly dependent. This can be modeled by saying that first order Markovianity is here generically defined on a different time ordering done by using a *Adaptive Time* (AT) ordering variable k. In particular, the event with a lower k index can be (probabilistically) the cause of the $k + 2$ (or $k + 1$ in case of consecutive not null events) event if it occurs within a certain maximum window time. Let us define such a window time I as *Maximum Influence Window* (MIW). This dependence can be captured by a DBN only if the time instant at which the consequence happens is left to vary within the range from 0 to I. By using MIW it is easy to see that two events $a_{m_{k-2}}^j$ and $a_{m_k}^j$ are directly dependent if $\sum_{n=m_{k-1}}^{m_k} x_n^j \Delta t \leq I$. Therefore, using random RLE sequences as a basis, i.e. sequences where both elements of the couple $\left(x_k^j, a_{m_k}^j\right)$ are random values, a DBN can be formed where nodes at same time instant in a generic k index time slice represent effectual events, $a_{m_k}^j$ and relative time of occurrence r_k^j with respect to previous events, i.e. their cause. As a consequence, direct dependencies between different sequences of events can be first captured and learned from RLE sequences considering links between couples of slices indexed with generic $k-1$ and k discrete indexes assumed to remain within the influence windows I while conditional dependencies of relative time occurrences of given event pairs can be captured within each slice k. This is equivalent to define a *Homogeneous Influence Window* (HIW) $I_k(.,.)$ working on successive couples $\left(x_k^j, a_{m_k}^j\right)$ of the RLE sequence $\Upsilon_j = R\left(\Psi_j\right)$ to express a sparse first-order Markovianity criterion over the time ordered event space. However, this criterion is not equivalent to a first order Markovianity in the time space, but to a competitive set of higher-order Markovianity models (up to a Ith order Markovianity) on the absolute time used to model lower level strata of DBNs related to proto/core super states, states and sensor observations.

Three cases can be individuated, as follows: the first when $r_k^j = 2$, the second $2 < r_k^j \leq I/\Delta t$, the third is $r_k^j > I/\Delta t$. $\mathbf{I} = I/\Delta t$ represents the maximum ordered events that $I_k(.,.)$ can contain.

The first one happens when of two non-null events consecutively happen: $a_{m_{k-1}}^j \neq \emptyset$ and $a_{m_k}^j \neq \emptyset$, $(m_k - m_{k-1}) = 2$. In this situation first-order Markovianity in the time space corresponding with sparse first-order Markovianity defined over the time ordered event space.

The second case happens when $a_{m_{k-2}}^j$ and $a_{m_k}^j$ are two consequent non-null events separated by a null event $a_{m_{k-1}}^j = \emptyset$ after a absolute time interval $r_k^j \Delta t$.

The third case is similar to the previous one but $r_k^j > \mathbf{I}$. In this case causality cannot be anymore assumed $a_{m_{k-2}}^j$ and $a_{m_k}^j$. This can be also interpreted by saying that no dependency among non null events into two slices at adjacent AT k indexes. In the following paragraphs this concept is exploited to define Run Length Encoding E-DBNs (RLE E-DBNs) and interaction models.

5.3 RLE E-DBN for Entity Dynamic Modeling

Having considered RLE encoding definition for events sequence and the notation described in the previous section, it is now possible to model RLCE-DBNs. Let us consider a discrete index k that represents a generic AT instant of first event A_m^j in a time window MIW starting at absolute time t_m sampled each Δt. The index m_k in a RLE sequence temporally aligned at $t_0 = T_0$ with the same sample time Δt can be obtained as follows:

$$m_k = \text{argmax}_y \, y : \sum\nolimits_{n=1}^{y} x_n^j \Delta t \leq t_m \tag{6}$$

It is now possible to fix $Y_j(k) = \left[Y_1^j(k), \ldots, Y_{V_j}^j(k) \right]$ where $Y_s^j(k)$ are binary random variables:

$$Y_s^j(k) = \begin{cases} 1 & \text{when } a^{j,s} = \alpha_{m_k}^j, a^{j,x} \in \mathcal{E}^j, s \in 1, \ldots, V_j \\ 0 & \text{otherwise} \end{cases} \tag{7}$$

$Y_j(k)$ represents a generic DBN slice indexed by k. From above concept it is possible situations have to be represented:

(1) Null event: $\alpha_{m_k}^j = \emptyset$, e.g. $Y_j(k) = [0 \quad 0 \quad 0 \quad .. \quad 0\,]$.
(2) Non-null event $\alpha_{m_k}^j \neq \emptyset$, e.g. $Y_j(k) = [0 \quad 1 \quad 0 \quad .. \quad 0\,]$, with $x = 2$.

where $Y_x^j(k) = 1$ for a given x and it can be shown that $t_{m_k = } t_m$.

5.3.1 Null Event

Considering a sequence of events: $\left\{\ldots, \alpha^j_{m_{k-1}}, \alpha^j_{m_k}, \alpha^j_{m_{k+1}}, \ldots\right\}$, where $\alpha^j_{m_{k-1}} \neq \emptyset$, $\alpha^j_{m_k} = \emptyset$, $\alpha^j_{m_{k+1}} \neq \emptyset$ and $\alpha^j_{m_{k-1}} \neq \alpha^j_{m_{k+1}}$. We can define r^j_{k+1} as the time that separates $\alpha^j_{m_{k-1}}$ from the $\alpha^j_{m_{k+1}}$:

$$r^j_{k+1} = r^j_{k+1}(m_{k-1}) + r^j_{k+1}(m_{k+1}). \tag{8}$$

where $r^j_{k+1}(m_{k-1})$ represents the time between $\alpha^j_{m_{k-1}}$ and $\alpha^j_{m_k}$, while $r^j_{k+1}(m_{k+1})$ is the time between $\alpha^j_{m_k}$ and $\alpha^j_{m_{k+1}}$. If $\alpha^j_{m_k} = \emptyset$ (null event) occurs at time t_m, i.e. between two non-null events there is a stable situation, the Eq. (9) shows that the stability period can be divided into two parts, one before t_m and one after t_m.

5.3.2 Non-Null Event

Now we consider another sequence of events: $\left\{\ldots, \alpha^j_{m_k}, \alpha^j_{m_{k+1}}, \ldots\right\}$, where $\alpha^j_{m_k} \neq \emptyset$, $\alpha^j_{m_{k+1}} \neq \emptyset$ and $\alpha^j_{m_k} \neq \alpha^j_{m_{k+1}}$. We can define r^j_{k+1} as the time that separates $\alpha^j_{m_k}$ from the $\alpha^j_{m_{k+1}}$:

$$r^j_{k+1} = r^j_{k+1}(m_{k+1}). \tag{9}$$

When two non-null events occur consecutively, r^j_{k+1} is smaller than the MIW.

It is possible to define a generic couple of homogenous non-null events as two DBN slices for k and $k + 1$ indexes (which can be either consecutive or not) as: $\left(Y^j_s(k), Y^j_i(k+1)\right)$. A vector of temporal variables $T^j_{i,s,t}(k+1)$ can be defined as:

$$T^j_{i,s,t}(k+1) = \begin{cases} 1 & \text{when } t = r^j_{k+1}(m_{k+1}) \in [0, I],\ \mathrm{a}^{j,s} = \alpha^j_{m_{k+1}},\ \mathrm{a}^{j,i} = \alpha^j_{m_k}\mathrm{a}^{j,s},\ \mathrm{a}^{j,i} \in \mathcal{E}^j, i, s \in 1, \ldots, V_j \\ 0 & \text{otherwise} \end{cases} \tag{10}$$

Vector $T^j_{i,s,t}(k+1)$ can be interpreted as binary random variable vector

$$T^j_{i,s,t}(k+1) = \left[T^j_{i,s,0}, \ldots, T^j_{i,s,\mathrm{r}}, \ldots, T^j_{i,s,\mathbf{I}}\right], \tag{11}$$

where $T^j_{i,s,\mathrm{r}}(k+1) = 1$ when the time at which event $\alpha^j_{m_{k+1}}$ occurs after event $\alpha^j_{m_k}$ is equal to r, i.e. $r = r^j_{k+1}(k+1)$. In this case it is possible to define non-zero conditional probability for the couple of events $\left(Y^j_s(k), Y^j_i(k+1)\right)$ as follows:

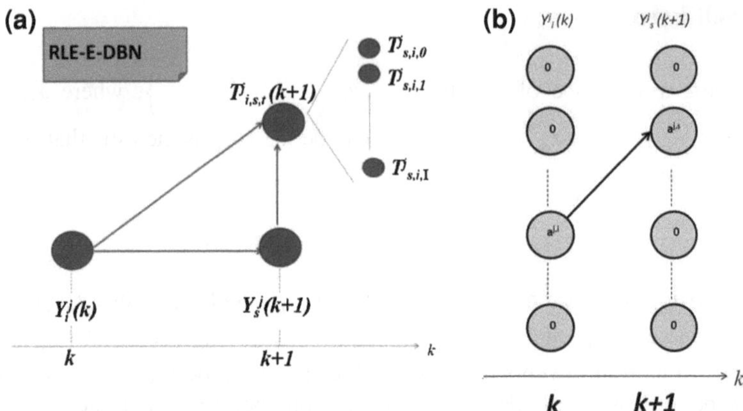

Fig. 5 Examples of: RLE E-DBN structure (**a**) and corresponding slices (**b**) for generic couple of events $\left(Y_s^j(k), Y_i^j(k+1)\right)$

$$p\left(Y_i^j(k+1) = 1 | Y_s^j(k) = 1\right). \tag{12}$$

The vector $T_{i,s,t}^j(k+1)$ just defined represents a second hierarchic level, in which the occurrence time between two events of the same entity is stored. In particular, the probability of $T_{i,s,r}^j(k+1)$ comes out to be dependent on which couple of successive events occurred at time k and $k+1$. In Fig. 5 a hierarchic RLE E-DBN structure is shown. In particular, the probability of $T_{i,s,t}^j(k+1)$ comes out to be dependent on which pair of successive events occurred at time k and $k+1$, i.e. it possible to write the conditioned probabilities as follows:

$$p\left(T_{i,s,t}^j(k+1) = 1 | Y_i^j(k+1) = 1, Y_s^j(k) = 1\right). \tag{13}$$

The structure depicted in Fig. 5 gives the possibility to calculate the joint probability of couples of consecutive (k-ordered) events time as follows:

$$p\left(Y_s^j(k), Y_i^j(k+1)\right) = p\left(Y_i^j(k+1)/Y_s^j(k)p\left(Y_s^j(k)\right)\right) \tag{14}$$

5.4 RL Coupled E-DBN for Interactive Entities Dynamic Modeling

Let us suppose one has to deal with two RLE sequences, Υ_P and Υ_C, which represent strings of event for different entity indicated as j and j'. Under the hypothesis that Υ_P and Υ_C, are produced by a time aligned couple of proto and

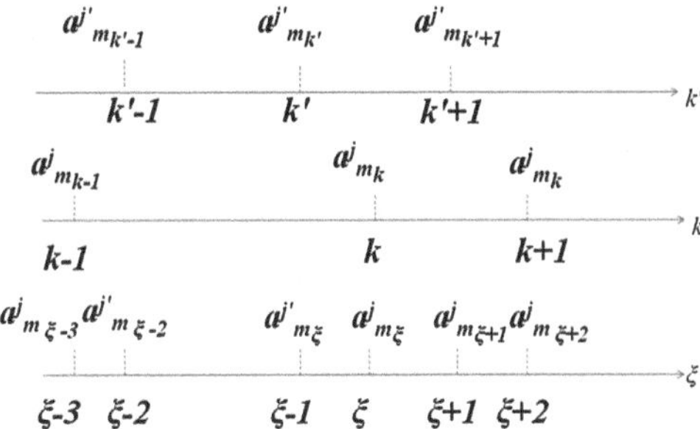

Fig. 6 RLE sequences for $j\,j'$ entities, IRLE built by time re-ordering

core intelligent sensors sharing the same starting time t_0 and processing data at the same time sampling Δt. Now, we explain previous concepts taking into consideration two time ordered sequences for separated entities, j and j'. By time re-ordering of j and j' strings, a sequence of non-homogenous events is generated, as shown in Fig. 6.

As done before, it is now possible to define their *RLE Influence* (I-RLE) sequence $\Upsilon_{jj'}$ in ordered time, as: $\Upsilon_{jj'} = \left\{\ldots\ldots!x^j_{\xi-1}\mathrm{a}^j_{m_{\xi-1}}!x^{j'}_{\xi}\mathrm{a}^{j'}_{m_\xi}!x^j_{\xi+1}\right.$ $\left.\mathrm{a}^j_{m_{\xi+1}}!\ldots\ldots\right\}$. Where only one event kind occurs between two homogeneous events, ξ represents the ordered index (similar to k used before). We can define a triplet of events as three slices, corresponding to two different DBNs (i.e., DBN for j entity and DBN for j' entity): $\left(Y^j_s(\xi-1), Y^{j'}_{s'}(\xi), Y^j_i(\xi+1)\right)$. Also in this case, in order to describe cause-effect relationships between two non-homogeneous events, it is possible to define a *Non-Homogeneous Influence window* $I'_\xi(.,.)$ working on successive couples $\left(x^{j'}_\xi, \mathrm{a}^{j'}_{m_\xi}\right)$ and $\left(x^j_{\xi+1}, \mathrm{a}^j_{m_{\xi+1}}\right)$ of the I-RLE non-homogeneous sequence $\Upsilon_{jj'} = R\left(\Psi_{jj'}\right)$. It is possible to define a *Maximum Influence Window between* no homogeneous events as I'.

Now we can show to possibly represent interactions between the middle, last and first events $\mathrm{a}^{j'}_{m_\xi}$, $\mathrm{a}^j_{m_{\xi+1}}$ and $\mathrm{a}^{j'}_{m_{\xi-1}}$ in I-RLE sequences, using RL Coupled E-DBN.

Example we consider, for simplicity, a specific event sequence as follows: $\left\{\ldots, \alpha^j_{m_{\xi-1}}, \alpha^{j'}_{m_\xi}, \alpha^j_{m_{\xi+1}}, \ldots\right\}$, where $\alpha^j_{m_{\xi-1}}, \alpha^{j'}_{m_\xi}, \alpha^j_{m_{\xi+1}} \neq \emptyset$ with $\alpha^j_{m_{\xi-1}} \neq \alpha^{j'}_{m_\xi} \neq \alpha^j_{m_{\xi+1}}$. In this situation first-order Markovianity in time space is equal to Markovianity in ξ-ordered time space.

We can divide the sequence in two sub-sequences of events, as follows: $\left\{\alpha^j_{m_{\xi-1}}, \alpha^{j'}_{m_\xi}\right\}$ and $\left\{\alpha^{j'}_{m_\xi}, \alpha^j_{m_{\xi+1}}\right\}$. According to the concepts described before (Sect. 5.3), it is possible to define two couples of non-homogeneous and non-null events as: $\left(Y^{j'}_{s'}(\xi), Y^j_s(\xi-1)\right)$ and $\left(Y^j_i(\xi+1), Y^{j'}_{s'}(\xi)\right)$, in which $j \neq j'$ and $i, s \in 1, \ldots, V_j$ while $s' \in 1, \ldots, V_{j'}$. For each couple of non-homogeneous events, a vector of temporal variables can be associated as follows:

$$\left(Y_i^j(\xi+1), Y_{s'}^{j'}(\xi)\right) \rightarrow T_{s',s,t}^{jj'}(\xi+1); \left(Y_{s'}^{j'}(\xi), Y_s^j(\xi-1)\right) \rightarrow T_{i,s',t}^{jj'}(\xi), \tag{15}$$

where the vectors $T_{s',s,t}^{jj'}(\xi+1)$ and $T_{i,s',t}^{jj'}(\xi)$ are defined as follows (they have the same form, for simplicity we will show only $T_{s',s,t}^{jj'}(k+1)$ definition):

$$T_{s',s,t}^{jj'}(\xi+1) = \begin{cases} 1 & \text{when } t = r_{\xi+1}^{jj'}(m_{\xi+1}) \in \left[0, I'\right], a^{j',s'} = \alpha_{m_{\xi+1}}^{j'}, a^{j,s} = \alpha_{m_\xi}^{j}, \\ 0 & \text{otherwise} \end{cases}$$

where $a^{j',s'} \in \mathcal{E}^{j'}, a^{j,s} \in \mathcal{E}^j; s' \in 1, \ldots, V_{j'}, s \in 1, \ldots, V_j$.

The vector $T_{s',s,t}^{jj'}(k+1)$ can be interpreted as binary random variable vector:

$$T_{s',s,t}^{jj'}(\xi+1) = \left[T_{s',s,0}^{jj'}, \ldots, T_{s',s,r'}^{jj'}, \ldots, T_{s',s,\mathbf{I}'}^{jj'}\right], \tag{16}$$

where $T_{s',s,r'}^{jj'}(\xi+1) = 1$ when the time at which event $\alpha_{m_{k+1}}^{j'}$ occurs after event $\alpha_{m_k}^{j}$ is equal to r', i.e. $r' = r_{k+1}^{jj'}(\xi+1)$.

Also in this case it is possible to define two probabilistic dependencies: $p\left(Y_{s'}^{j'}(\xi)/Y_s^j(\xi-1)\right)$ and $p\left(Y_i^j(\xi+1)/Y_{s'}^{j'}(\xi)\right)$ because the events occur into I'.

Considering the sequence of the triplet, shown before, $\left\{\ldots, \alpha_{m_{\xi-1}}^j, \alpha_{m_\xi}^{j'}, \alpha_{m_{\xi+1}}^j, \ldots\right\}$, we assume the following notations: $\mathcal{E}_-^j \triangleq \alpha_{m_{\xi-1}}^j$, $\mathcal{E}^{j'} \triangleq \alpha_{m_\xi}^{j'}$, $\mathcal{E}_+^j \triangleq \alpha_{m_{\xi+1}}^j$.

Under the hypothesis that $I = I' + I'$, which corresponds to $\mathbf{I} = \mathbf{I}' + \mathbf{I}'$, it possible to write the conditional probabilities of $\mathcal{E}_+^j$ given $\mathcal{E}^{j'}$ and $\mathcal{E}_+^j$, as $p\left(\mathcal{E}_+^j/\mathcal{E}^{j'}, \mathcal{E}_-^j\right)$.

The joint probabilities of the triplet of events $p\left(\mathcal{E}_-^j, \mathcal{E}^{j'}, \mathcal{E}_+^j\right)$ are describe by RLCE-DBNs and mathematically defined by Bayes theorem as follows:

$$p\left(\mathcal{E}_-^j, \mathcal{E}^{j'}, \mathcal{E}_+^j\right) = p\left(\mathcal{E}_+^j/\mathcal{E}^{j'}, \mathcal{E}_-^j\right)p\left(\mathcal{E}^{j'}/\mathcal{E}_-^j\right)p\left(\mathcal{E}_-^j\right) \tag{17}$$

The occurrence times are embedded into a more compact structure which can be seen as an upper hierarchic level represented by a temporal histogram. Two kinds of conditional probabilities $p\left(T_{i,s',t}^{jj'}(\xi+1) = 1|\mathcal{E}_-^j, \mathcal{E}^{j'}, \mathcal{E}_+^j\right)$ $p\left(T_{s',s,t}^{jj'}(\xi) = 1|\mathcal{E}_-^j, \mathcal{E}^{j'}, \mathcal{E}_+^j\right)$ are evaluated in order to detect when the event $\mathcal{E}_+^j$ takes place after $\mathcal{E}^{j'}$. It is possible to assume the following definitions: passive interactions are referred to $(\mathcal{E}_-^P, \mathcal{E}^C, \mathcal{E}_+^P)$ triplets (proto-core-proto), while active interactions are associated to $(\mathcal{E}_-^C, \mathcal{E}^P, \mathcal{E}_+^C)$ (core-proto-core). In Fig. 7 a hierarchic RLCE-DBN structure is shown.

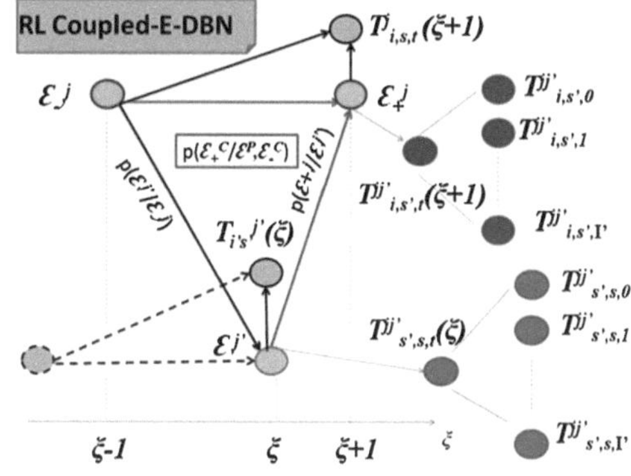

Fig. 7 Example of a RLCE-DBN for generic triplet of events $\left(\mathcal{E}_{-}^{j}, \mathcal{E}^{j'}, \mathcal{E}_{+}^{j}\right)$ where it is possible to note three hierarchic levels in which are stored the occurrence time between events

5.5 Switching Model for Interaction Representation

In the previous section a probabilistic model based on RLCE-DBNs was described, in order to compress information on interactions.

The RLCE-DBNs are hierarchical structures: this makes it possible to describe the relationships between two entities at different resolution levels. The proposed framework is composed by two layers: state transition model layer (low-level) and a so called Influence Model layer (IM) (high-level).

The IM (high-level layer) permits to establish whether a triplet corresponds to an autobiographical memory. Example, if an operator observes normal crowding situations, the related actions stored into AM will be significantly different from the actions performed by another operator. The state transition model (low-level layer) is basically a HMM describing state transitions. Each layer corresponds to a probabilistic model. These models are linked to higher or lower levels by a switching variable. The RLCE-DBNs switching model is able to describe interaction information from a macroscopic (event transitions) to a microscopic (state transitions) point of view, at the same time.

Considering a generic triplet of events, $\left\{\mathcal{E}_{-}^{j}, \mathcal{E}^{j'}, \mathcal{E}_{+}^{j}\right\}$, it can be referred to an influence model (e.g., a specific AM). We can then define a switching variable θ as influence parameter, Fig. 8. The joint probability of the triplets and of switching variable is:

$$p\left(\mathcal{E}_{-}^{j}, \mathcal{E}^{j'}, \mathcal{E}_{+}^{j}, \theta\right) = p\left(\mathcal{E}_{+}^{j} / \mathcal{E}^{j'} \mathcal{E}_{-}^{j}, \theta\right) p\left(\mathcal{E}^{j'} / \mathcal{E}_{-}^{j}, \theta\right) p\left(\mathcal{E}_{-}^{j} / \theta\right) p(\theta) \tag{18}$$

For switching parameter estimation the posterior density is computed, as follows:

$$p\left(\theta / \mathcal{E}_{-}^{j} \mathcal{E}^{j'} \mathcal{E}_{+}^{j}\right) = \frac{p\left(\mathcal{E}_{-}^{j} \mathcal{E}^{j'} \mathcal{E}_{+}^{j} / \theta\right) p(\theta)}{p\left(\mathcal{E}_{-}^{j} \mathcal{E}^{j'} \mathcal{E}_{+}^{j}\right)}. \tag{19}$$

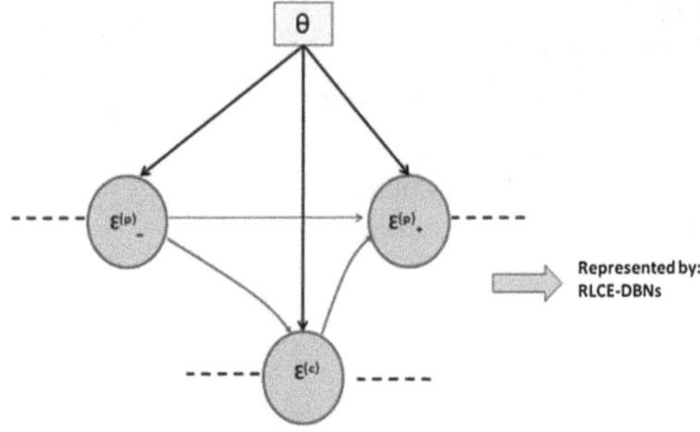

Fig. 8 Example of RLCE-DBNs for passive triplet, e.g. $\{\mathcal{E}_{-}^{P}, \mathcal{E}^{C}, \mathcal{E}_{+}^{P}\}$, with a parameter θ tied across proto-core-proto transitions

The triplet of events, embodied into C RLE E-DBNs, can be considered as elementary unit of relationships because causes and effects are contained. The AM, in some cases, can be used not only to learn (e.g. off line learning), but also to classify interactions (e.g. on line prediction).

To perform prediction the task, we consider a passive triplet $\{\mathcal{E}_{-}^{P}, \mathcal{E}^{C}, \mathcal{E}_{+}^{P}\}$: when an external event $\mathcal{E}^{C}$ is detected by the system, the proto map is analyzed to select the previously occurred internal event $\mathcal{E}_{-}^{P}$. The autobiographical memory is then examined to establish which internal event $\hat{\mathcal{E}}_{+}^{P}$ is the more likely:

$$\hat{\mathcal{E}}_{+}^{P} : \max_{\hat{\mathcal{E}}_{+}^{P}, x_{\xi+1}, x_{\xi-1}} P\left(\frac{\mathcal{E}_{+}^{j}, \varnothing^{(x_{\xi+1})}}{\begin{array}{c}, \\ \mathcal{E}^{j'}, \varnothing^{(x_{\xi-1})}, \mathcal{E}_{-}^{j}\end{array}}\right). \tag{20}$$

Equation (19) provides a way to predict internal events which are more likely to occur, considering the time at which the $\hat{\mathcal{E}}_{+}^{P}$ might take place.

We explained (Sect. 5.1) that each event is defined as a label change: $\mathcal{E}^{j} \triangleq (\sigma_{m}^{j}, \sigma_{m'}^{j})$ if $\sigma_{m}^{j} \neq \sigma_{m'}^{j}$ where $\sigma_{m}^{j}, \sigma_{m'}^{j} \in \bar{L}_{j}$ with $m, m' = 1, \ldots, M$, M is number of labels. σ^{j} represents the switching variable that describes states-space transition model into Super states, whose network is given in Fig. 9.

The RLCE-DBN is able to anticipate not only future events, but also to provide a state space transition model. This can be represented by a first-order Markov model [36]. The RLCE-DBN integrates different levels into the same multi-hierarchical structure, in which the lower resolution level (high level layer) describes event transitions, while a high granularity resolution (low level layer) accounts for state transition model. This permits a refined prediction of event and state changing.

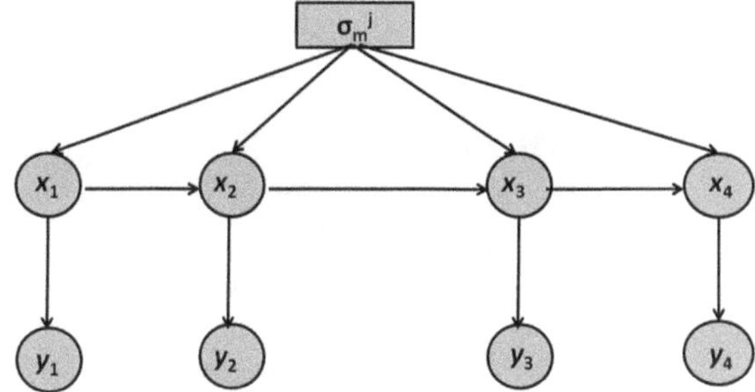

Fig. 9 State space model with tied parameter on the latent transitions

6 Applications on Crowd Behavior Analysis

The theory developed so far has been applied to a simulated crowd monitoring scenario.

A simulator was developed in order to gather data for validation of the interaction theory presented in Sect. 5. Such data, involving interacting crowd and human operator, is not available in public datasets and in any case not easily collectible in a real scenario. A well balanced integration of realism in the behavior of the crowd and in the visualization of it was here needed, at least to some extent as already mentioned. The crowd within the simulator was modeled according to state of the art approaches reviewed in Sect. 2. A cognitive framework was implemented, as proposed in Sect. 3. Interactions between the crowd and a monitoring entity have been modeled according to theory developed in Sect. 5, after extracting information from the simulator's output as depicted in Sect. 4.

6.1 The Simulator

The simulated monitored environment is shown in Fig. 10. The configuration of doors, walls and rooms is however customizable and a wide range of scenarios can be set for tests.

The use of a graphical engine (freely available at http://www.horde3d.org/) has been introduced in order to make the simulation realistic in the autobiographical memory (Sect. 4.2) training phase. Here a human operator acts on doors configuration in order to prevent room overcrowding, based on the visual output, which need to be as realistic as possible. Namely, the simulator has to output realistic data both from the behavioral point of view, in order to effectively interact with the human operator, and from the visual point of view, in order to grant an effective interface by truly depicting reality. Reactions of an operator faced with an unrealistic visual output could be extremely different and strongly depend on rendering quality. For this reason, characters are also animated to simulate walk motion

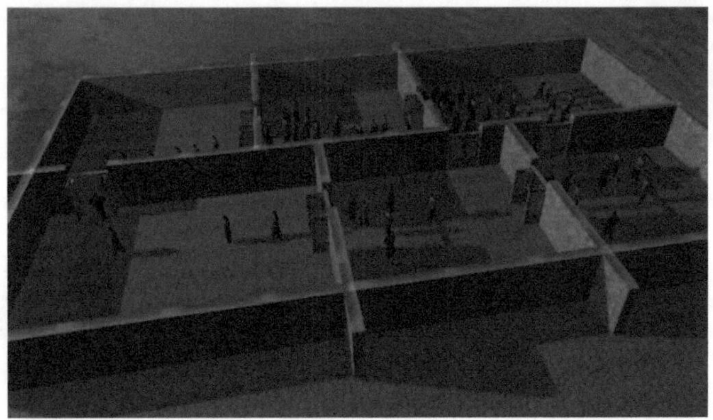

Fig. 10 The simulated monitored environment

(at first glance a crowded environment with still people could look less populated than it really is).

Crowd behavior within the simulator is modeled based on social forces, which were mentioned in Sect. 2. This model assimilates each character on the scene to a particle subject to 2D forces, and threats it consequently from a strictly physical point of view. Its motion equations are derived from Newton's law $\vec{F} = m\vec{a}$.

The forces a character is driven by are substantially of three kinds [28]. An attractive motivational force $\vec{F}_{mot}$ pulls characters toward some scheduled destination, while repulsive physical forces $\vec{F}_{phy}$ and interaction forces $\vec{F}_{int}$ prevent from collision into physical objects and take into account interactions within characters. An additional linear drag (viscous resistance) $\vec{F}_{res}$ takes into account the fact that no character actually persists in its state of constant speed but tends to stop its motion as motivation runs out. This force is in fact accounted for and included in $\vec{F}_{mot}$. The three forces are estimated at each time instant for each character, whose position is then updated according to the motion equation and normalized according to the current fps rate supported by the graphical engine (which strongly depends on the number of characters to be handled).

People incoming rate is modeled as a Poisson distribution. Their "death" occurs as they get to their final scheduled destination. A human operator interacts with the crowd by opening doors to let it flow, while trying to minimize the time a doors remains open. Although somehow simplified with respect to [28] (where additional assumptions on trajectories' regularity are made) the developed model results in a good overall output, where people behave correctly.

The simulator also includes (simulated) sensors. These try to reproduce (processed) sensor data coming from different cameras looking at different subsets (rooms) of the monitored scene. A virtual people estimation algorithm outputs the number of people by simply adding some noise to the mere number of people framed by the virtual cameras. These can be placed anywhere in the rooms, with virtually any position, angle and zoom parameters (Fig. 11). Video sequences can

Fig. 11 Different camera views

be recorded from the simulator and real VS algorithms could be exploited for people monitoring and count estimation. Future researches will head this why and try to apply actual state-of-the-art systems for people counting. This would give additional consistency to this work.

6.2 Training and Knowledge Storage

As already mentioned in Sect. 3, and again stressed through Sects. 4 and 5, the designed system is able to extract information during a learning phase, in order to represent complex interactions between the dynamic observed situation and operators' actions.

Such a learning phase actually includes two conceptually separated steps.

In a first step, the system learns how to indeed extract information from the environment and organize it. This phase corresponds to the training of the SOM (Sect. 4.2) which is in charge of clusterizing core feature vectors, thus defining Super-States. Here the system autonomously learns similarities and shapes its own way of associating similar objects [32]. The way reality is represented by the system obviously raises issues on all the subsequent steps, in which such information is used and processed, but we will not go in details here. We just mention that the way SOMs classify input vectors can strongly depend on the nature of the input vectors the SOM was fed with during the training phase. Even the ordering of such input training vectors can influence the shaping of the neural network's connections. Eventually, the dimension of the SOM (which determines the number of labels to be associated to Super-States) cannot be optimized a priori.

Here a 10×10 2D SOM is trained by feeding it with input vectors coming from several simulations. As explained in Sect. 4.2, this means there are 100 possible core Super-States $S_{i,C}$, which define 100^2 possible core events (including 100 possible null core events). On the other hand, seven doors connect the rooms. These can be either open or close. The proto status is then represented as a seven dimensional binary vector. This defines 2^7 possible door configurations proto, namely Super-States $S_{i,P}$ and thus 128^2 possible core events (again including null events).

The second learning step consists in learning interactions i.e. learning connections within and between the DBNs the model is composed by. This was shown in deep details in Sect. 5. The system stores such knowledge in two autobiographical memories, namely a passive and an active memory. The active memory allows for prediction: given a triggering proto event, it can give probabilities of the most likely core (external) event to occur. On the other hand, the passive memory allows for a closed-loop automatic reaction of the system: given a triggering core event (e.g. some specific crowding situation), the memory supplies the more likely reaction (which can be a non-reaction as well!) of human operator and, yet again, can predict its more likely consequences by switching back to the active memory. Roughly speaking, this allows for moving the acquired knowledge to the decision block and "closing the switch" in Fig. 4.

It must be pointed out that the proposed approach has some limitations, namely, it cannot deal directly with situations which were never observed during the training phase. An AM does not have the capability (as opposite to a SOM) of managing situations that are just similar (and not identical) to its learned experience. This fact underlines the importance of a good training phase for the AM to work properly, but also of a fine tuning of the parameters which regulate information extraction. As already mentioned, the way interactions are modeled is not only affected by the parameters of the model itself, such as causality influence windows, but also by the nature of processed data available. Future developments of the theory may aim at handling such shortcomings. For this very reason, in any case, the proposed system always gives a human operator some control, namely the chance to step in the decision block to compensate for shortcomings of the AM or at least to monitor it, as shown in Fig. 4.

6.3 Prediction and Inference

Figure 12 shows a fragment of an active AM: event triplets are stored together with the probabilities $p\left(\mathcal{E}_{+}^{j}/\mathcal{E}^{j'},\mathcal{E}_{-}^{j}\right)$ (Eq. 16) and temporal histograms (Eq. 15) referred to the event $\mathcal{E}_{+}^{C}$ with respect to the triggering $\mathcal{E}^{P}$. Such information allows the system to predict what will be the most likely reaction of the environment to some proto action (event) and when the former is most likely to occur. In this case, the following situation is depicted: after the crowd remains for a more than some fixed time (namely I', defined in Sect. 5.4, which is here taken to be equals to 12 s) in a Super-State $S_{i,C}$ (whose label is 7), a null core event is detected. By changing the door configuration from 0100000 to 1100000 (proto event) two equally likely events can occur: either the crowd persists in its Super-State (null event) or the

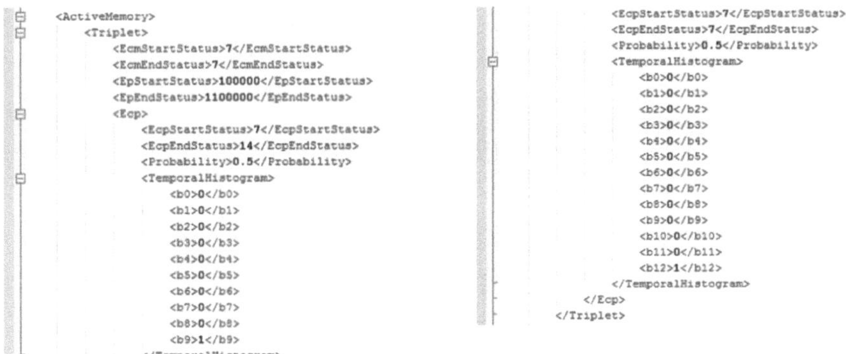

Fig. 12 Fragment of active AM (xml file)

crowd can shift to the Super-State labeled as 14. The corresponding RLCE-DBN diagram is depicted in Fig. 13 (cfr. Figure 7).

At the same time, a passive memory can work in a closed loop to act in place of the human operator. Given a triggering core event $\mathcal{E}^C$, the AM can be searched to extract knowledge on which, and how likely, possible operator's reactions (i.e. proto events $\mathcal{E}^P_+$) are, given the former proto event $\mathcal{E}^P_-$ which has been causing $\mathcal{E}^C$. The system can then select the most likely human reaction and autonomously decide and act. In this sense we have been implementing knowledge transfer.

This closed-loop automatic setup has been tested on a simulated scenario. At least at a macroscopic level the AM is able to predict and avoid overcrowded situations acting to some extent as the training human operator. Unknown events sequences, which cannot be handled by the AM, are just reported. Decision and action are then commissioned to the guardian, which can in any case bypass the automatic system at any time.

Moreover, according to the concept of switch, introduced at the beginning of Sect. 5.5, an enhanced system was developed, where a bank (just two in our case) of AM were trained by different operators. These can show different ways of interacting with the environment, and thus construct different interaction models, which can better (or worse) describe reality depending on the observed evolution. The switching variable θ, which was previously introduced, allows for switching between different evolution representations, namely AMs, whenever one of them better describes observed proto-core interactions. The switching occurs at IM layer level as already pointed out. Performances of such a switching-based system look as good as the simpler non-enhanced structure. We expect better performances in testing more complicated situations. Previous studies on multiple-AM systems for automatic learning and human behavior analysis and classification [31] strongly persuade us to expect so.

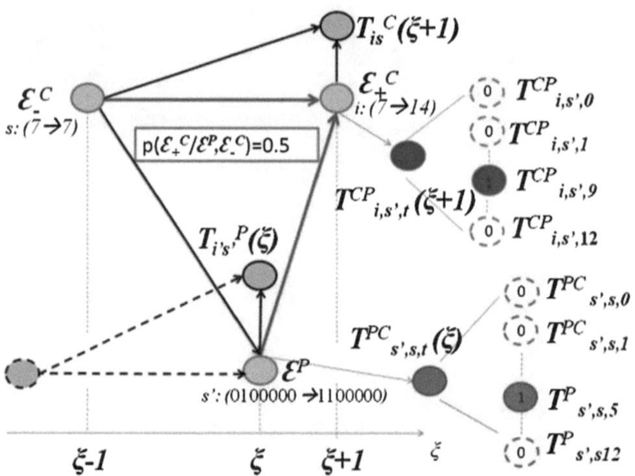

Fig. 13 RLCE-DBN representation for active triplet

7 Conclusions and Future Works

In this work a crowd monitoring application was presented, where a CN embedding E-DBN knowledge can interact with a visual simulator of crowd situations. Such a bio-inspired model was applied to define causal relationship between internal and external entities and a simulation platform was developed to provide a large set of training data, compressed by a RL encoding. Knowledge gathered from such event training set was stored within one or more AMs, which allows for making prediction and inference at decision block level in the CN.

Future steps of this work will closer investigate the enhanced switching-based model by testing it on more complicated scenarios. Moreover, the way feature reduction (by means of SOMs) influences knowledge representation will also be investigated. Eventually, an analysis on the impact of different SOM and AM training on prediction and inference will be carried out.

References

1. Remagnino, P., Velastin, S.A., Foresti, G.L., Trivedi, M.: Novel concepts and challenges for the next generation of video surveillance systems. Mach. Vis. Appl. **18**(3), 135–137 (2007)
2. Trivedi, M., et al.: Intelligent environments and active camera networks. In: Proceedings of the IEEE International Conference on System, Man and Cybernetics, vol. 2, pp. 804–809 (2000)
3. Lipton, A., Heartwell, C., Haering, N., Madden, D.: Automated video protection, monitoring & detection. IEEE Aerosp. Electron. Syst. Mag. **18**(5), 3–18 (2003)
4. Trivedi, M.M., et al.: Looking-in and looking-out of a vehicle: computer-vision-based enhanced vehicle safety. IEEE Trans. Intell. Transp. Syst. **8**(1), 108–120 (2007)

5. Damasio, A.R.: The Feeling of What Happens-Body, Emotion and the Making of Consciousness. Harvest Books, New York (2000)
6. Valera, M., Velastin, S.: Intelligent distributed surveillance systems: a review. IEEE Proc. Vis. Image Signal Process. **52**(2), 192–204 (2005)
7. Foresti, G.L., Regazzoni, C.S., Varshney, P.K.: Multisensor Surveillance Systems: The Fusion Perspective. Kluwer Academic, Boston (2003)
8. Collins, R., Lipton, A., Fujiyoshi, H., Kanade, T.: Algorithms for cooperative multisensory surveillance. Proc. IEEE **89**(10), 1456–1477 (2001)
9. Smith, D., Singh, S.: Approaches to multisensor data fusion in target tracking: a survey. IEEE Trans. Knowl. Data Eng. **18**(12), 1696–1710 (2006)
10. Prati, A., et al.: An integrated multi-modal sensor network for video surveillance. In: Proceedings of the Third ACM International Workshop on Video Surveillance & Sensor Networks (2005)
11. Chang, B.R., Tsai, H.F., Young, C.-P.: Intelligent data fusion system for predicting vehicle collision warning using vision/gps sensing. Expert. Syst. Appl. **37**(3), 2439–2450 (2010)
12. Wu, S., Decker, S., Chang, P., Camus, T., Eledath, J.: Collision sensing by stereo vision and radar sensor fusion. IEEE Trans. Intell. Transp. Syst. **10**(4), 606–614 (2009)
13. Zhan, B., Monekosso, D.N., Remagnino, P., Velastin, S.A., Xu, L.-Q.: Crowd analysis: a survey. Mach. Vis. Appl. **19**, 345–357 (2008)
14. Loscos, C., Marchal, D., Meyer, A.: Intuitive crowd behavior in dense urban environments using local laws. In: Proceedings of the Theory and Practice of Computer Graphics, pp. 122–129 (2003)
15. Liu, B., Liu, Z., Hong, Y.: A simulation based on emotions model for virtual human crowds. In: Fifth International Conference on Image and Graphics ICIG'09, pp. 836–840 (2009)
16. Handford, D., Rogers, A.: Modelling driver interdependent behavior in agent-based traffic simulations for disaster management. In: The Ninth International Conference on Practical Applications of Agents and Multi-Agent Systems, pp. 163–172 (2011)
17. Davies, A.C., Yin, J.H., Velastin, S.A.: Crowd monitoring using image processing. Electron. Commun. Eng. J. **7**, 37–47 (1995)
18. Wren, C.R., Azarbayejani, A., Darrell, T., Pentland, A.P.: Pfinder: real-time tracking of the human body. IEEE Trans. Pattern Anal. Mach. Intell. **19**, 780–785 (1997)
19. Haritaoglu, I., Harwood, D., David, L.S.: W4: real-time surveillance of people and their activities. IEEE Trans. Pattern Anal. Mach. Intell. **22**, 809–830 (2000)
20. Marana, A.N., Velastin, S.A., Costa, L.F., Lotufo, R.A.: Automatic estimation of crowd density using texture. Safety Sci. **28**(3), pp. 165–175 (1998)
21. Zhao, T., Nevatia, R.: Bayesian human segmentation in crowded situations. In: IEEE Computer Society Conference on Computer Vision and Pattern Recognition, vol. 2, p. 459 (2003)
22. Andrade, E., Blunsden, S., Fisher, R.: Hidden markov models for optical flow analysis in crowds. In: 18th International Conference on Pattern Recognition.(ICPR), vol. 1, pp. 460–463 (2006)
23. Benabbas, Y., Ihaddadene, N., Djeraba, C.: Motion pattern extraction and event detection for automatic visual surveillance. EURASIP J. Image Video Process. **2011**, 15 (2011)
24. Rahmalan, H., Nixon, M., Carter, J.: On crowd density estimation for surveillance. In: The Institution of Engineering and Technology Conference on Crime and Security, pp. 540–545 (2006)
25. Cupillard, F., Avanzi, A., Bremond, F., Thonnat, M.: Video understanding for metro surveillance. In: IEEE International Conference on Networking, Sensing and Control, vol. 1, pp. 186–191 (2004)
26. Mehran, R., Oyama, A., Shah, M.: Abnormal crowd behavior detection using social force model. In: IEEE conference on computer vision and pattern recognition CVPR 2009, pp. 935–942 (2009)

27. Pellegrini, S., Ess, A., Schindler, K., Gool, L. van.: You'll never walk alone: modeling social behavior for multi-target tracking. In: International Conference on Computer Vision, Miami, 20–26 June 2009
28. Luber, M., Stork, J.A., Tipaldi, G.D., Arras, K.O.: People tracking with human motion predictions from social forces. In: Proceedings of the International Conference on Robotics & Automation (ICRA), Anchorage (2010)
29. Moore, B.E., Ali, S., Mehran, R., Shah, M.: Visual crowd surveillance through a hydrodynamics lens. Commun. ACM **54**(12), 64–73 (2011)
30. Dore, A., Regazzoni, C.S.: Bayesian bio-inspired model for learning interactive trajectories. In: Proceedings of the IEEE International Conference on Advanced Video and Signal Based Surveillance, AVSS 2009, Genoa, Italy, 2–4 Sept 2009
31. Dore, A., Cattoni, A., Regazzoni, C.: Interaction modeling and prediction in smart spaces: a bio-inspired approach based on autobiographical memory. In: IEEE Transactions on Systems, Man and Cybernetics, Part A: Systems and Humans (2010)
32. Kohonen, T.: The self-organizing map. Proc. IEEE **78**(9), 1464–1480 (1990)
33. Dore, A., Soto, M., Regazzoni, C.S.: Bayesian tracking for video analytics. Signal Process. Mag. IEEE. **27**(5), pp.46–55, (2010) doi: 10.1109/MSP.2010.937395
34. Patnaik, D., Laxman, S., Ramakrishnan, N.: Discovering excitatory networks from discrete event streams with applications to neuronal spike train analysis. In: ICDM, pp. 407–416 (2009)
35. Oliver, N.M., Rosario, B., Pentland, A.P.: A bayesian computer vision system for modeling human interactions. IEEE Trans. Pattern Anal. Mach. Intell. **22**(8), 831–843 (2000)
36. Pan, W, Dong, W, Cebrian, M., Kim, T., Fowler, J.H., Pentland, A.S.: Modeling dynamical influence in human interaction: using data to make better inferences about influence within social systems. IEEE Signal Process. Mag. **29**(2), 77–86 (2012)

Unified Face Representation for Individual Recognition in Surveillance Videos

Le An, Bir Bhanu and Songfan Yang

Abstract Recognizing faces in surveillance videos becomes difficult due to the poor quality of the probe data in terms of resolution, noise, blurriness, and varying lighting conditions. In addition, the poses in the probe data are usually not frontal view, as opposed to the standard format of the gallery data. The discrepancy between the two types of data makes the existing recognition algorithm far less accurate in real-world surveillance video data captured in a multi-camera network. In this chapter, we propose a multi-camera video based face recognition framework using a novel image representation called Unified Face Image (UFI), which is synthesized from multiple camera video feeds. Within a temporal window the probe frames from different cameras are warped towards a template frontal face and then averaged. The generated UFI representation is a frontal view of the subject that incorporates information from different cameras. Face super-resolution can also be achieved, if desired. We use SIFT flow as a high level alignment tool to warp the faces. Experimental results show that by using the unified face image representation, the recognition performance is better than the result of any single camera. The proposed framework can be adapted to any multi-camera video based face recognition using any face feature descriptors and classifiers.

Keywords Face recognition · Face registration in video · Multi-camera network · Surveillance videos

---

L. An (✉) · B. Bhanu · S. Yang
Center for Research in Intelligent Systems, University of California,
Riverside, CA 92521, USA
e-mail: le.an@email.ucr.edu

B. Bhanu
e-mail: bhanu@cris.ucr.edu

S. Yang
e-mail: songfan.yang@email.ucr.edu

Augment Vis Real (2014) 6: 123–136
DOI: 10.1007/8612_2012_2

Published Online: 31 August 2012

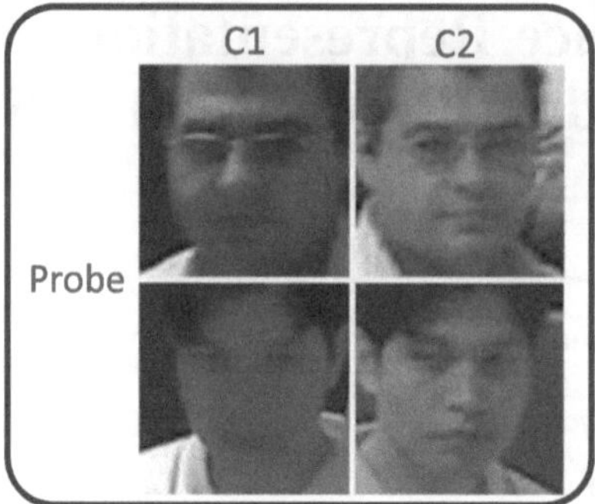

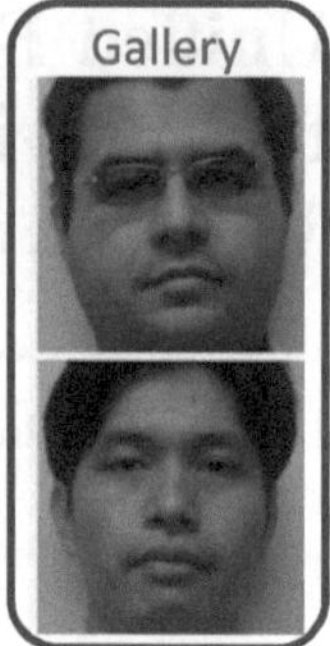

Fig. 1 Sample data from the ChokePoint dataset [21]. The subjects' faces were captured by multiple cameras. The appearance between the probe data and the gallery data is considerably different

1 Introduction

With the wide deployment of surveillance video cameras, the necessity to perform robust face recognition in surveillance videos is rising for the purpose of access control, security monitoring, etc. Although face recognition has been studied extensively, it is still very challenging for the existing face recognition algorithms to work accurately in real-world surveillance data. Empirical studies have shown that the face image of approximate size 64×64 is required for existing algorithms to yield good results [13]. However, when a subject is not in the close vicinity of the camera, the captured face would have very low resolution. In addition, video sequences often suffer from motion blur and noise, together with changes in pose, lighting condition and facial expression. With the low resolution face images captured by surveillance cameras in different lighting conditions and poses, the recognition rate could drop dramatically to less than 10 % as reported in [6].

The challenge of face recognition in surveillance video is mainly due to the uncontrolled image acquisition process with the non-cooperative subject. The subject is often moving and it is not uncommon that only non-frontal view is captured, while in the gallery set often frontal view is stored. With multiple cameras in the surveillance system, each camera is likely to capture the face from different viewpoints. Fig. 1 shows sample probe data from 2 cameras (C1 and C2) and gallery data in the ChokePoint dataset [21]. Note that the appearance of the probe data is significantly different from the gallery data.

How to tackle the discrepancy between the probe and gallery data becomes critical in developing a robust recognition algorithm. In addition, how to utilize video inputs from multiple cameras to improve the recognition performance is also an essential part for face recognition in surveillance camera systems. In this chapter, a new image based face representation generated from multiple camera inputs is proposed to improve the recognition accuracy in real-world multi-camera surveillance data.

The rest of this chapter is organized as follows. Related works and our contributions are outlined in Sect. 2. Technical details are provided in Sect. 3. Sect. 4 shows the experimental results and conclusions are made in Sect. 5.

2 Related Work and Contributions

For face recognition in video, a rough categorization divides different methods into two classes: 3D model based face recognition and 2D image based face recognition. Here we provide some pointers to the representative work.

To tackle the modality mismatch between the probe and the gallery data, a strategy is to build a 3D face model to handle varying poses. In [5] a 3D morphable model was generated as a linear combination of basis exemplars. The model was fit to an input image by changing the shape and albedo parameters of the model. The drawback of the 3D based approach is the high computational cost. Zhang and Samaras [23] combined spherical harmonics illumination representation with 3D morphable models. Aggarwal and Harguess [7] used average-half-face instead of the whole face to improve the face recognition accuracy for 3D faces. Barreto and Li [10] proposed a framework for 3D face recognition system with variation of expression. The disadvantage for 3D based recognition is the high computational cost in building the 3D model. In addition, constructing a 3D model from low-resolution inputs is very difficult when the facial control points cannot be accurately localized by detectors.

To cope with the low-resolution issue in video based face recognition, Hennings-Yeomans et al. [9] used features from the face and super-resolution priors to extract a high-resolution template that simultaneously fits the super-resolution and face feature constraints. A generative model was developed in [3] for separating the illumination and down-sampling effects to match a face in a low-resolution video sequence against a set of high resolution gallery sequences. Stallkamp et al. [16] introduced a weighting scheme to evaluate individual contribution of each frame in a video sequence. In [14] face images with different modalities are projected into a common subspace for matching. Recently, Biswas et al. [4] proposed a learning-based likelihood measurement to match high-resolution gallery images with probe images from surveillance videos. The performance of these methods generally degrades when applied to real-world surveillance data. In addition, the learning based methods may not be viable due to the insufficient training data that are available in reality.

Additional effort has been made to recognize faces from different input resources. Normally a face captured from a single camera contains information of partial face only. To overcome this limitation, some approaches have been proposed that use multiple cameras to improve the recognition performance. A cylinder head model was built in [8] to first track and then fuse face recognition from multiple cameras. In [22] a reliability measure was trained and used to select the most reliable camera for recognition. A two-view face recognition was proposed in [19] where the recognition results are fused using the Bayesian based approach. However, these approaches were validated only on videos of much higher resolution compared to the real-world surveillance data.

As a surveillance system often consists of multiple cameras, the multi-camera based face recognition approach is naturally desired. In this chapter, we propose a

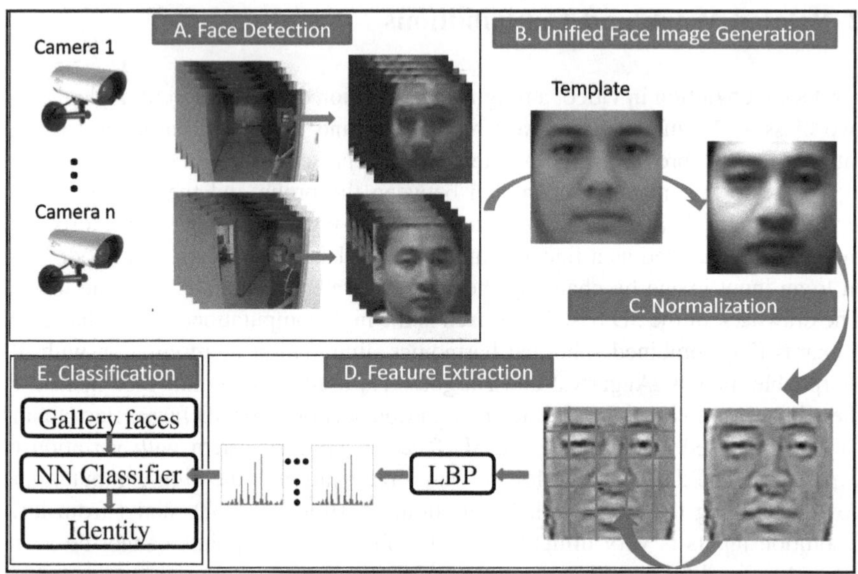

Fig. 2 System overview. After the faces are extracted from the video sequence, the UFIs are generated. Lighting effects is normalized before the feature extraction and classification

framework for multi-camera video based face recognition by generating a new face image representation called Unified Face Image (UFI). From a set of multi-camera probe videos, a UFI is generated using several consecutive frames from each camera. These frames are first warped towards a frontal face template and the warped images are then averaged to obtain the UFI. SIFT flow [3] is used to warp the images. Given probe sequences from multiple cameras, only a few UFIs are needed to be extracted. The fusion is performed at the image level and the appearance of the generated UFIs is more coherent with the gallery data. The proposed framework can be used in any video based face recognition algorithms using different feature descriptors, classifiers or weighting schemes.

3 Technical Details

In this section the proposed framework for face recognition in surveillance videos is presented. Fig. 2 gives an outline of the system.

After the face images are extracted from the video sequences, the UFIs are generated by fusing these face images. A frontal view face template is used to warp the face images. The warping is achieved using SIFT flow. Before the classification, the generated UFIs are lighting normalized to elliminate the shading effects and non-uniform lighting conditions. Local Binary Pattern (LBP) is employeed as the face descriptor. Before we describe UFI generation process, SIFT flow as the warping method is introduced.

3.1 SIFT Flow

SIFT flow was recently reported in [3] as an effective way to align images at the scene level. SIFT flow is a dense matching algorithm that uses SIFT features [12] to find the pixel-to-pixel correspondences between two images. It is shown in [3] that scene pairs with high complexity can be robustly aligned. In the first step, SIFT features for every pixel are extracted. Then similar to optical flow, an energy function is minimized to match two images s_1 and s_2 :

$$E(\mathbf{w}) = \sum_{\mathbf{p}} min(\|s_1(\mathbf{p}) - s_2(\mathbf{p} + \mathbf{w}(\mathbf{p}))\|_1, t) \tag{1}$$

$$+\sum_{\mathbf{p}} \eta(|u(\mathbf{p})| + |v(\mathbf{p})|) \tag{2}$$

$$+\sum_{(\mathbf{p},\mathbf{q})\in\varepsilon} min(\alpha|u(\mathbf{p}) - u(\mathbf{q})|, d) + min(\alpha|v(\mathbf{p}) - v(\mathbf{q})|, d) \tag{3}$$

where $\mathbf{p}$ is the image grid. $\mathbf{w}(\mathbf{p}) = (u(\mathbf{p}), v(\mathbf{p}))$ is the flow vector in horizontal and vertical direction at $\mathbf{p}$. ϵ defines a local neighborhood (a four-neighborhood is used). The data term in (1) is a SIFT descriptor match constraint that enforces the match along the flow vector $\mathbf{w}(\mathbf{p})$. (2) is the small displacement constraint that ensures the flow vector $\mathbf{w}(\mathbf{p})$ to be as small as possible without additional information. The smoothness constraint is imposed in (3) for the pixels in the local neighborhood. t and d are the thresholds for outliers and flow discontinuities. η and α are the scaling factors for the small displacement and smoothness constraint. In this energy function the truncated $L1$ norm is used in the data term and the smoothness term.

The dual-layer loopy belief propagation is used in the optimization process. Then, a coarse-to-fine SIFT flow matching scheme is applied to improve the matching result and the computation efficiency [3].

Fig. 3 illustrate the face alignment by SIFT flow. The input image is aligned towards the frontal view template. As can be seen, although the input images are not frontal face, the output image after alignment is very close to the frontal view. D1 shows the absolute difference between the template and the input image. D2 shows the absolute difference between the template and the aligned image. The difference in D2 is much smaller compared to the difference in D1 due to pose variation.

3.2 Unified Face Image (UFI) Generation

After being extracted from the original sequence, the faces are used to generate the UFI. The face captured by the surveillance cameras are often not frontal view. Direct matching the non-frontal faces to the frontal view gallery data often lead to

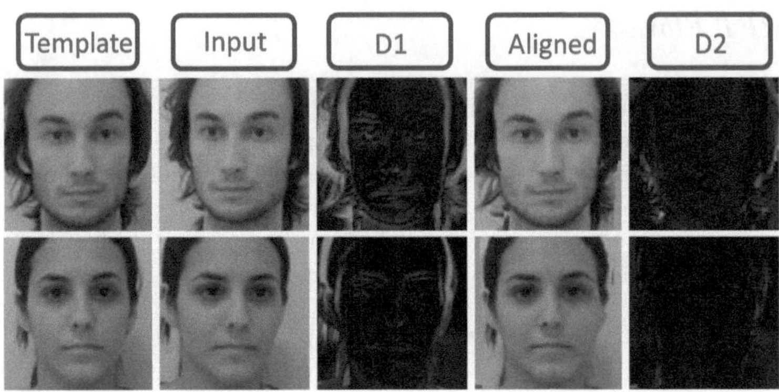

Fig. 3 Using SIFT flow to align the input image with the template. D1 shows the difference between the template and the input image, D2 shows the difference between the template and the aligned image. After SIFT flow alignment, the output image is very similar to the template image

Fig. 4 Sample data from the FEI dataset [4]

poor recognition results. To overcome this limitation, we warp the face images towards a common face template. The template I_0 is obtained by averaging the aligned frontal faces in the ChokePoint and the FEI datasets [4] with 225 subjects in total. By using the average face as the template, we avoid warping the face towards any specific subject. Fig. 5 shows some sample data from the FEI dataset (Fig. 4).

In a temporal window centered at time t, the UFI is generated as

$$UFI(t) = \frac{1}{(2k+1)C} \sum_{i=-k}^{k} \sum_{j=1}^{C} \langle I_j(t+i), I_0 \rangle \tag{4}$$

where $I_j(t+i)$ is the frame at time $t+i$ from camera j. C is the total number of cameras and $2k+1$ is the length of the temporal window. $\langle I_j(t+i), I_0 \rangle$ warps $I_j(t+i)$ towards the template I_0 using SIFT flow. Since different cameras have different field-of-view, the information from each frame is complementary to each other. The averaging is essentially an information fusion process to aggregate all

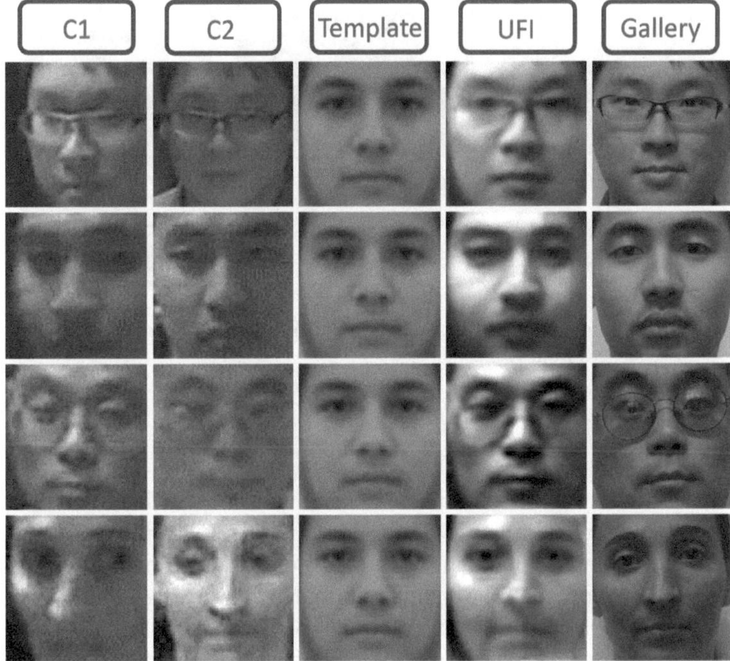

Fig. 5 UFI is generated from two camera inputs using the warping template. C1 and C2 shows one of the frames in each camera used to generate the UFI. The UFI is the frontal view of the subjects which are very similar to the gallery data

the information from different frames at different views. The generated UFI is a concise representation for all of the $(2k+1)C$ frames. Fig. 5 shows some samples of the generated UFIs using faces from two cameras (C1 and C2).

As can be seen in Fig. 5, the generated UFIs are the frontal views of the subjects. The UFIs have less deviation from the gallery data in appearance. During this warping-averaging process, the noise and blurriness are suppressed and the facial details are enhanced. The UFI in the next temporal window is generated in the same manner.

For a given set of video sequences from multiple cameras, the number of generated UFIs depends on the number of total frames in each sequence and the time step between two UFIs, given that the sequences from different cameras have the same length. Fig. 6 is a sample illustration of how the UFIs are generated from a set of sequences. The generated UFIs have similar appearance due to the overlap between the temporal windows, the small difference between frames, and the alignment process in which all the faces are aligned towards the same template. When the overlap between the adjacent temporal windows decreases or the video is captured at a lower frame rate, the variation between the UFIs would increase.

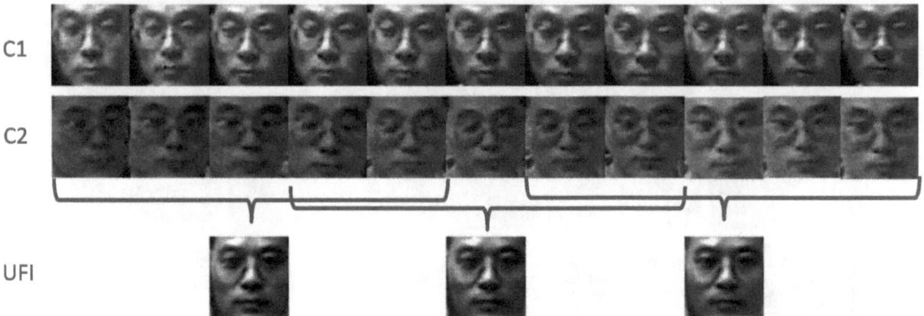

Fig. 6 A sample illustration of UFI generation in video sequences. Several UFIs are generated using a set of sequences from two cameras

3.3 Recognition

Since the UFIs are generated from data of different cameras, the different lighting conditions in the original frames will introduce non-uniform lighting in UFIs (see Fig. 5). In order to reduce the lighting effects, we use the normalization method in [7] to preprocess the UFIs. The lighting normalization includes four steps. In the first step the Gamma correction is performed to enhance the local dynamic range of the image. The second step involves Difference of Gaussian (DOG) filtering which has a bandpass behavior to eliminate to shading effects. Then, the facial regions that are either irrelevant or too variable are masked out. In the final step the contrast equalization is applied to scale the image intensities. After the lighting normalization, the non-uniform lighting effects are eliminated. The faces in the gallery are processed similarly. Fig. 7 shows some sample results by lighting normalization.

After the lighting normalization, we extract features from UFIs to match with the gallery image. There have been various face descriptors developed to encode the micro-pattern of face, such as local binary patterns (LBP) [1], local phase quantization (LPQ) [2], Gabor wavelets [15], patterns of oriented edge magnitudes (POEM) [20], etc. We choose LBP as the face descriptor for its simplicity. The face image is divided into blocks and the LBP features are extracted from each block. The final feature vector of the face is obtained by concatenating these block features. Note that in the proposed framework any feature descriptors can be adopted. Fig. 8 shows the basic LBP operator. The intensity values of the neighbor pixels are compared to the center pixel. A binary code is generated and its decimal value is used for histogram binning.

We apply a nearest-neighbor (NN) classifier. The Chi-square distance is used to compute the feature distance as suggested in [1]. The Chi-square distance between two histograms M and N is computed by

$$\chi^2(M,N) = \sum_i \frac{(M_i - N_i)^2}{M_i + N_i} \tag{5}$$

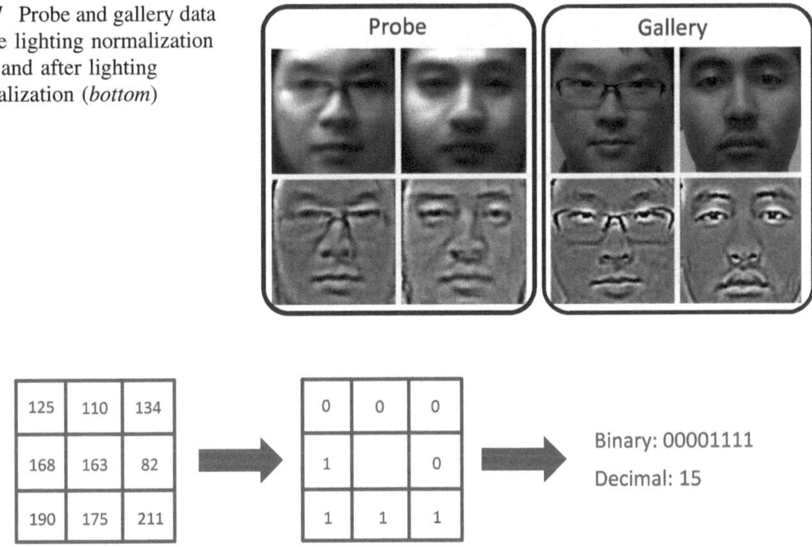

Fig. 7 Probe and gallery data before lighting normalization (*top*) and after lighting normalization (*bottom*)

Fig. 8 Basic LBP descriptor generated from a 3 × 3 image patch

The distance scores are accumulated for all the UFIs generated from the original set of sequences and the lowest summed score across all the gallery images provides the identity of the subject. Each UFI is considered equally important yet any frame weighting scheme [16] can be applied to the UFIs to further improve the recognition performance.

4 Experiments

4.1 Dataset and Settings

We use the ChokePoint dataset [21] which is designed for evaluating face recognition algorithms under real-world surveillance conditions. A subset of the video sequences from portal 1 (P1) in two directions (Entering and Leaving) and two cameras (C1 and C2) are used (P1E_S1_C1, P1E_S1_C2, P1E_S2_C1, P1E_S2_C2, P1E_S3_C1, P1E_S3_C2, P1E_S4_C1, P1E_S4_C2, P1L_S1_C1, P1L_S1_C2, P1L_S2 _C1, P1L_S2_C2, P1L_S3_C1, P1L_S3_C2, P1L_S4_C1, P1L_S4_C2). In total 25 subjects are involved. The gallery set contains the high-resolution frontal faces of the 25 subjects. In addition, a subset of the video sequences from portal 2 (P2) in one direction (Leaving) and two cameras (C1 and C2) are used (P2L_S1_C1, P2L_S1_C2, P2L_S2_C1, P2L_S2_C2, P2L_S3_C1, P2L_S3_C2, P2L_S4_C1, P2L_S4_C2). We do not use images from P2E due to insufficient data. In portal 2 there are 29 subjects. The extracted faces are provided with the dataset.

Table 1 Rank-1 recognition rates for the testing sequences in portal 1

	P1E_S1 (%)	P1E_S2 (%)	P1E_S3 (%)	P1E_S4 (%)	P1L_S1 (%)	P1L_S2 (%)	P1L_S3 (%)	P1L_S4 (%)	Average (%)
C1	12	16	12	16	32	44	36	44	26.5
C2	8	12	12	16	32	12	32	28	19
UFI	44	48	32	40	40	40	48	56	43.5

Table 2 Rank-1 recognition rates for the testing sequences in portal 2

	P2L_S1 (%)	P2L_S2 (%)	P2L_S3 (%)	P2L_S4 (%)	Average (%)
C1	37.93	17.24	6.9	20.69	20.69
C2	31.03	41.38	10.34	44.83	31.90
UFI	79.31	55.17	37.93	68.97	60.35

The probe faces are normalized to 64×64. For each sequence, the initial 20 frames are chosen to form a challenging problem where the subjects were far away from the cameras. To generate UFI at the current frame, its previous and future 4 frames and itself are used (when the previous or future frame are not available, its mirror image with respect to the current frame is used, e.g., $I(t+1)$ is used when $I(t-1)$ is not available). In our method, we use 4 UFIs generated from the 20 frames at every fifth frame. We use the default parameters as provided in the implementation of [7] to normalize the lighting effects. $LBP^{u2}_{8,2}$ is used as suggested in [1]. The image block size is chosen as 16×16.

4.2 Experimental Results

To focus on the recognition improvement using UFIs generated from multiple camera data, we compare the results to the baseline method where each original probe frame in a single camera is used to match with the gallery images. The distance score for each frame is summed across the 20 frames in the sequence and the final identity is taken as the one with the lowest total score. We do not directly compare with the results on the ChokePoint dataset reported in [21] where a *video-to-video verification* protocol is used. The *video-to-image recognition* in our case is more challenging due to the significant data discrepancy between the probe and the gallery data.

Table 1 shows the rank-1 recognition rates in portal 1 and Table 2 shows the rank-1 recognition rates in portal 2. Compared to the recognition rates from individual cameras, the proposed new face representation improves the recognition rate remarkably in all but one set of the testing sequences (P1L_S2). On average, the recognition rate using UFI is 17 % higher than the result by camera 1 and 24.5 % higher than the result by camera 2 in portal 1, and the improvements rise to 39.66 and 28.45 % respectively in portal 2. The reason for the improved

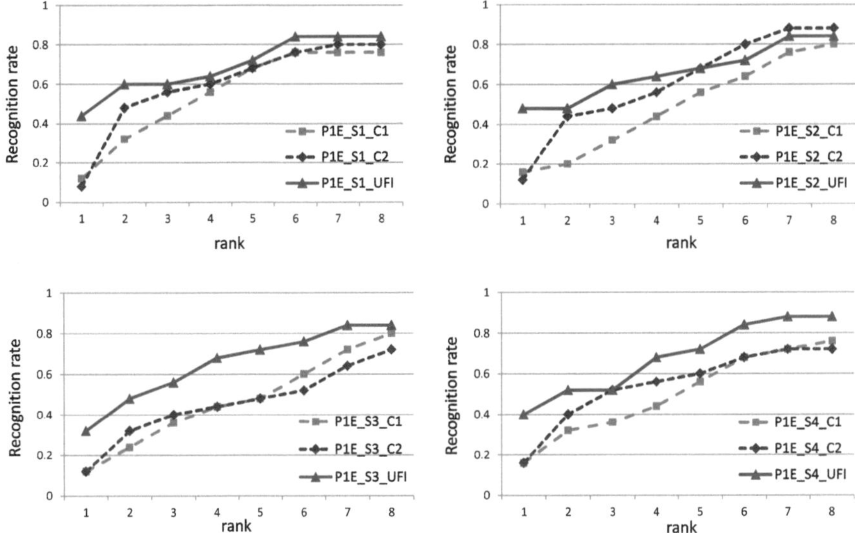

Fig. 9 Cumulative match curves for the testing sequences in portal 1, when the subjects are entering the portal (P1E)

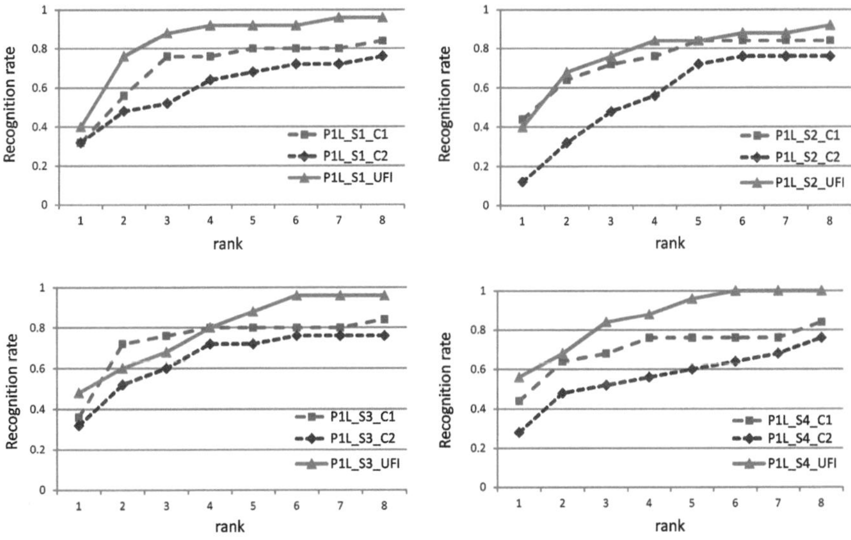

Fig. 10 Cumulative match curves for the testing sequences in portal 1, when the subjects are leaving the portal (P1L)

recognition performance is that by using UFIs as the new probe data, the discrepancy between the appearance of the probe data and the gallery data is reduced. By fusing the information from two cameras, the recognition result is superior to a single camera.

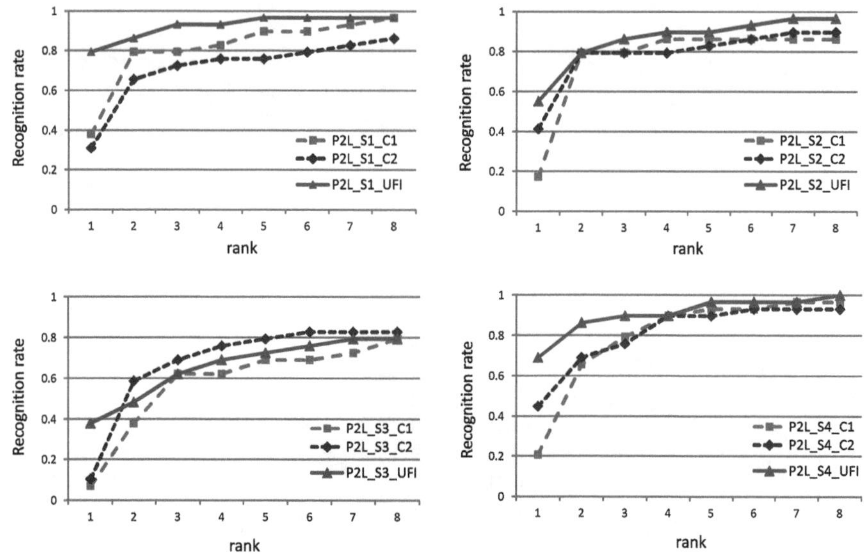

Fig. 11 Cumulative match curves for the testing sequences in portal 2, when the subjects are leaving the portal (P2L)

The cumulative match curves (CMC) are given in Figs. 9, 10 and 11 for sequences in P1E, P1L and P2L respectively. In general the recognition rates at different ranks are higher by using the proposed UFIs which congregate the useful face information from different cameras. The fusion achieved at the image level enables the easy adoption of different feature descriptors or classifiers. Moreover, no training or complex modeling is required.

5 Conclusions

A significant challenge for face recognition from surveillance videos is the mismatch between the frontal view gallery data and diverse appearance in the probe data. In this chapter, to overcome this limitation and to utilize the information from multiple cameras, we propose a novel image representation called Unified Face Image (UFI) by fusing the face images from different cameras. The face images are aligned towards a frontal view template using SIFT flow algorithm. The generated UFI is the frontal view of the subject. In this way the complementary information from multiple cameras is effectively combined. Given multiple video sequences as inputs, a few UFIs are generated for the subsequent recognition purpose. The experimental results on a public surveillance video based dataset indicate that by using the UFIs, the recognition rate is significantly improved compared to the recognition result from any single camera. The proposed method is simple yet effective and any feature descriptors, weighting schemes or classifiers can be easily adopted in this framework.

References

1. Ahonen, T., Hadid, A., Pietikäinen, M.: Face recognition with local binary patterns. In: European Conference on Computer Vision, pp. 469–481 (2004)
2. Ahonen, T., Rahtu, E., Ojansivu, V., Heikkila, J.: Recognition of blurred faces using local phase quantization. In: 19th International Conference on Pattern recognition 2008. ICPR 2008, pp. 1–4 (2008). doi:10.1109/ICPR.2008.4761847
3. Arandjelović, O., Cipolla, R.: A manifold approach to face recognition from low quality video across illumination and pose using implicit super-resolution. In: International Conference on Computer Vision (2007)
4. Biswas, S., Aggarwal, G., Flynn, P.: Face recognition in low-resolution videos using learning-based likelihood measurement model. In:International Joint Conference on Biometrics (IJCB) 2011, pp. 1–7 (2011). doi:10.1109/IJCB.2011.6117514
5. Blanz, V., Vetter, T.: Face recognition based on fitting a 3D morphable model. IEEE Trans. Pattern Anal Mach. Intell. **25**(9), 1063–1074 (2003). doi:10.1109/TPAMI.2003.1227983
6. Grgic, M., Delac, K., Grgic, S.: SCface—surveillance cameras face database. Multimedia Tools Appl. **51**(3), 863–879 (2011). Doi: 10.1007/s11042-009-0417-2. http://dx.doi.org/10.1007/s11042-009-0417-2
7. Harguess, J., Aggarwal, J.: A case for the average-half-face in 2D and 3D for face recognition. In: Computer Vision and Pattern Recognition Workshops, 2009. IEEE Computer Society Conference on CVPR Workshops 2009, pp. 7–12 (2009). doi:10.1109/CVPRW.2009.5204304
8. Harguess, J., Hu, C., Aggarwal, J.: Fusing face recognition from multiple cameras. In: Workshop on Applications of Computer Vision (WACV) 2009, pp. 1–7 (2009). doi:10.1109/WACV.2009.5403055
9. Hennings-Yeomans, P., Baker, S., Kumar, B.: Simultaneous super-resolution and feature extraction for recognition of low-resolution faces. In: Computer Vision and Pattern Recognition, 2008. IEEE Conference on CVPR 2008, pp. 1–8 (2008). doi:10.1109/CVPR.2008.4587810
10. Li, C., Barreto, A.: An integrated 3D face-expression recognition approach. In: Acoustics, Speech and Signal Processing, 2006. ICASSP 2006 Proceedings. 2006 IEEE International Conference on, vol. 3, p. III (2006). doi:10.1109/ICASSP.2006.1660858
11. Liu, C., Yuen, J., Torralba, A.: SIFT flow: Dense correspondence across scenes and its applications. IEEE Trans. Pattern Anal. Mach. Intell. **33**(5), 978–994 (2011). doi:10.1109/TPAMI.2010.147
12. Lowe, D.G.: Distinctive image features from scale-invariant keypoints. Int. J. Comput. Vis. **60**(2), 91–110 (2004). doi:10.1023/B:VISI.0000029664.99615.94. url:http://dx.doi.org/10.1023/B:VISI.0000029664.99615.94
13. Lui, Y.M., Bolme, D., Draper, B., Beveridge, J., Givens, G., Phillips, P.: A meta-analysis of face recognition covariates. In: Biometrics: Theory, Applications, and Systems, 2009. BTAS '09. IEEE 3rd International Conference on, pp. 1 –8 (2009). doi:10.1109/BTAS.2009.5339025
14. Sharma, A., Jacobs, D.: Bypassing synthesis: PLS for face recognition with pose, low-resolution and sketch. In: Computer Vision and Pattern Recognition (CVPR), 2011 IEEE Conference on, pp. 593–600 (2011). doi:10.1109/CVPR.2011.5995350
15. Shen, L., Bai, L.: A review on gabor wavelets for face recognition. Pattern Anal. Appl. **9**(2), 273–292 (2006). doi:10.1007/s10044-006-0033-y. url:http://dx.doi.org/10.1007/s10044-006-0033-y
16. Stallkamp, J., Ekenel, H., Stiefelhagen, R.: Video-based face recognition on real-world data. In: Computer Vision, 2007. ICCV 2007. IEEE 11th International Conference on, pp. 1–8 (2007). doi:10.1109/ICCV.2007.4408868
17. Tan, X., Triggs, B.: Enhanced local texture feature sets for face recognition under difficult lighting conditions. Image Proces. IEEE Trans. **19**(6), 1635–1650 (2010). doi:10.1109/TIP.2010.2042645

18. Thomaz, C.E., Giraldi, G.A.: A new ranking method for principal components analysis and its application to face image analysis. Image Vis. Comput. **28**(6), 902–913 (2010). doi:10.1016/j.imavis.2009.11.005. url:http://www.sciencedirect.com/science/article/pii/S0262885609002613
19. Tsai, G.Y., Tang, A.W.: Two-view face recognition using bayesian fusion. In: Systems, Man and Cybernetics, 2009. SMC 2009. IEEE International Conference on, pp. 157–162 (2009). doi:10.1109/ICSMC.2009.5346579
20. Vu, N.S., Caplier, A.: Enhanced patterns of oriented edge magnitudes for face recognition and image matching. Image Proces IEEE Trans. **21**(3), 1352–1365 (2012). doi:10.1109/TIP.2011.2166974
21. Wong, Y., Chen, S., Mau, S., Sanderson, C., Lovell, B.: Patch-based probabilistic image quality assessment for face selection and improved video-based face recognition. In: Computer Vision and Pattern Recognition Workshops (CVPRW), 2011 IEEE Computer Society Conference on, pp. 74–81 (2011). doi:10.1109/CVPRW.2011.5981881
22. Xie, B., Ramesh, V., Zhu, Y., Boult, T.: On channel reliability measure training for multi-camera face recognition. In: Applications of Computer Vision, 2007. WACV '07. IEEE Workshop on, p. 41 (2007). doi:10.1109/WACV.2007.46
23. Zhang, L., Samaras, D.: Face recognition from a single training image under arbitrary unknown lighting using spherical harmonics. Pattern Anal. Mach. Intell. IEEE Trans. **28**(3), 351–363 (2006). doi:10.1109/TPAMI.2006.53

Person Re-identification in Wide Area Camera Networks

Shishir K. Shah and Apurva Bedagkar-Gala

Abstract Person re-identification is defined as the problem of matching images or videos of people taken from different cameras. It is a fundamental task in wide area surveillance for multi-camera tracking and the subsequent analysis of long term activities and behaviors of people in the scene. A person's appearance is most often used to characterize their identity and is matched across cameras to establish re-identification. In typical surveillance scenarios, people are observed from a distance, do not present similar views across cameras, and the environment is uncontrolled and varying, which makes extracting, learning, and matching a person based on their appearance a non-trivial task. Person re-identification has been an area of intense research in the past five years, nonetheless, it remains an open problem and many of its aspects and challenges are not addressed. In this chapter, we explore the problem of person re-identification in wide area camera networks, in the context of consistent tracking over multiple cameras in order to facilitate the estimation of the global trajectory of a person over the camera network. We discuss the problem details, identify the requirements of a reasonable re-identification model, and highlight the challenges. We also present a multi-parametric model, demonstrate its effectiveness for person re-identification, and discuss its contributions.

Keywords Re-identification · Appearance model · Open set matching · Closed set matching

S. K. Shah (✉) · A. Bedagkar-Gala
Deptartment of Computer Science, University of Houston, 4800 Calhoun, 501 PGH, Houston, TX 77204-3010, USA
e-mail: sshah@central.uh.edu

Augment Vis Real (2014) 6: 137–162

DOI: 10.1007/8612_2012_7

Published Online: 5 September 2012

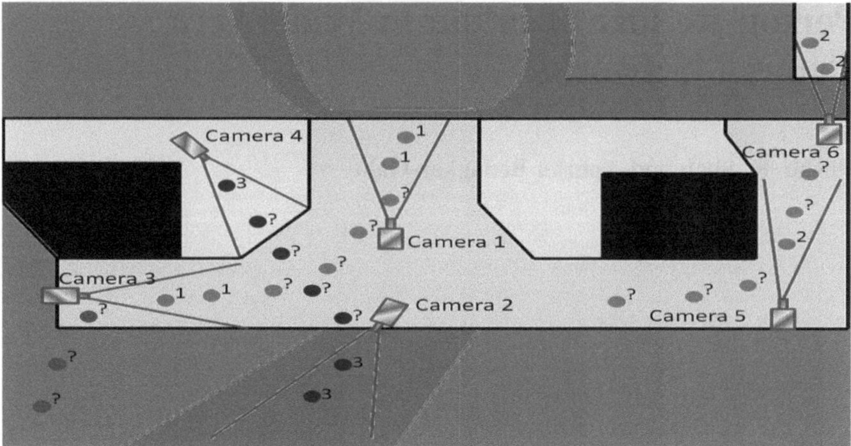

Fig. 1 Illustration of re-identification in the context of wide area camera networks

1 Introduction

Network of cameras with non-overlapping fields-of-views (FOVs) are often employed to provide enhanced visibility of large geospatial areas. Such networks are increasingly used for a variety of applications ranging from surveillance to understanding of traffic flow and space usage. A fundamental task across such applications is tracking of people across multiple cameras. Consistent tracking of people over multiple cameras requires the ability to re-identify a person as he/she leaves the FOV of one camera and enters the FOV of another. Person re-identification is defined as the process of matching images or videos of a person taken from two different cameras. Apart from the domain of surveillance, re-identification is pervasive in other applications like photo tagging [1] and photo browsing. Re-identification models can be also used for surveillance applications within a single camera, for example, to determine if a person visits a shop multiple times or if the same person or another person picks up a bag left by someone at a specific location. Hence, the objective for person re-identification is to assign a consistent label to multiple observations of a person, thereby improving the semantic coherence of analysis.

Figure 1 shows an example of a camera network wherein the position of several cameras are shown overlaid on the floor plan of an office building to illustrate the relative placement of cameras. The colored dots represent people seen by specific cameras within the network with their "IDs" displayed beside the dots. As person denoted by "ID = 1" (represented by green dot) leaves the FOV of camera 1 and enters the FOV of camera 3, re-identification enables the reacquisition of the ID to facilitate the association of disconnected tracks generated by two cameras.

In order to maintain a consistent ID for a person over space and time, any observation of a person needs to be searched across all previous observations from

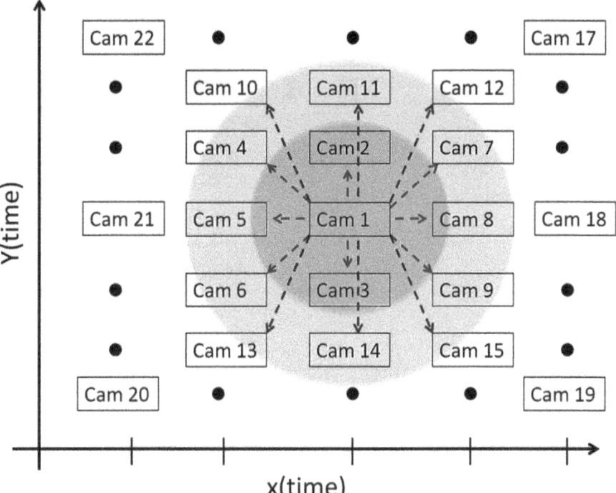

Fig. 2 The network of cameras placed on a grid illustrating the spatio-temporal relationships between them

other cameras within the network. In the absence of prior knowledge about scene geometry and/or camera placements, the person ID has to be searched between every camera pair in the network. If the spatial neighbors of a camera are known or can be determined, the re-identification camera set can be constrained and the search minimized.

To illustrate the most general case of re-identification, consider the cameras in a network viewed within a spatial grid as a function of time as shown in Fig. 2. Let us assume that a person is observed for the first time in "Cam1". For all succeeding observations in all cameras, re-identification needs to be established between observations from "Cam1" (we will refer to it as the gallery camera) and the all other network cameras (we will refer to these as the probe set cameras). As the initially observed person progresses through the camera network, the probe set is updated to cameras further away. Thus, the probe set is changing as a function of space. In addition, as the re-identification progresses, the gallery set is continually expanded to include the most recent set of probe cameras.

Consider the specific case of re-identification across a single set of camera pair at each time step as shown in Fig. 3. This would be the case when the spatial neighbor of a camera is known and along with the direction in which people are moving. The first time a person is observed, his/her appearance model is learned, and the subject is enrolled in the gallery set. Hence, all people observed in the second camera would form the probe set. After re-identification, all the people observed in the second camera that were previously unseen are enrolled into the gallery. As the re-identification moves to the next camera pair the gallery set is extended.

Hence, person re-identification in the context of wide area camera networks is an open set matching problem where the gallery evolves over time. Further, the probe set dynamically changes for each camera pair, and all the probes within a set are not necessarily a subset of the gallery. In addition, there might be several

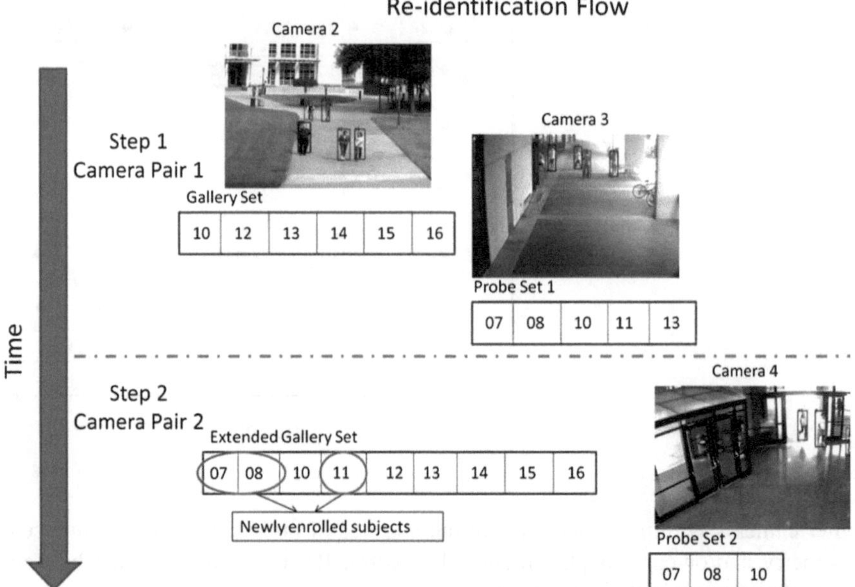

Fig. 3 Multiple person re-identification as open set matching

subjects that co-exist in time and need to be re-identified simultaneously. Thus, re-identification is not a single person but a multiple-person matching problem.

2 Challenges

Re-identification is a rather challenging problem and has been the subject of intense research in the past five years [2–6]. The primary challenges are attributed to variations in appearance, viewing angles, and articulation of the human form. Figure 4 shows images of a person taken from four different cameras that highlight some of these challenges. The appearance of a person can be different between two camera views due to different illumination conditions. The observed pose and scale of a person may be different due to varying distances and viewing angles between the camera and the person.

In typical scenarios, people movements are unconstrained and the environments are uncontrolled. This makes extracting and learning a person's appearance challenging. In addition, sensor resolution and placement might not provide a good quality image needed to extract usable information. People can also be partially or completely occluded due to crowds or clutter. Inconsistencies in detection and tracking results can also negatively impact the process of re-identification. Biometrics like face and gait are difficult to extract and utilize due to frame rate and

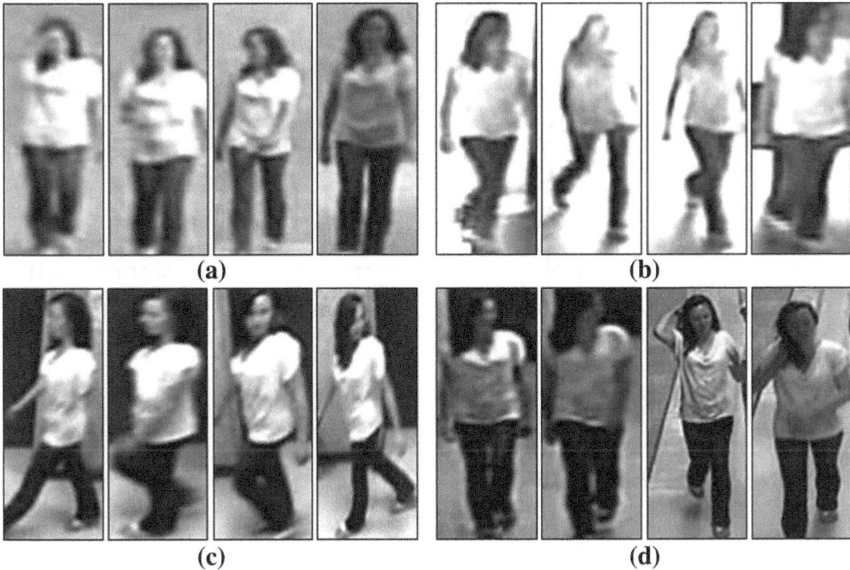

Fig. 4 Images of a person taken from four different cameras to illustrate appearance variations

sensor resolution constraints. Hence, appearance is by far the most readily available cue that can be used. Nonetheless, appearance based models can breakdown and can be non-specific to a person's appearance. They are useful if the re-identification is over short intervals but can become increasingly uncertain as the time between observations increases. Thus, there is need for developing and extracting a reliable and unique person model that can prove useful for the re-identification problem.

3 Related Work and Open Issues

Appearance models are most often used for re-identification and spatio-temporal relationships between cameras are used to reason about false matches. Traditionally, the focus has been to develop appearance models that are discriminative and robust to viewpoint, pose and illumination variations. The primary assumption here is that people do not change their clothing over the observation period. Such models have been tested primarily in closed set experiments, where the gallery is fixed and re-identification is established on a single probe assuming that the probe is always present in the gallery.

Most of the work on person re-identification can be classified into two categories: *non-learning* approaches and *learning* approaches. Figure 5 summarizes the relevant taxonomy of current work in person re-identification.

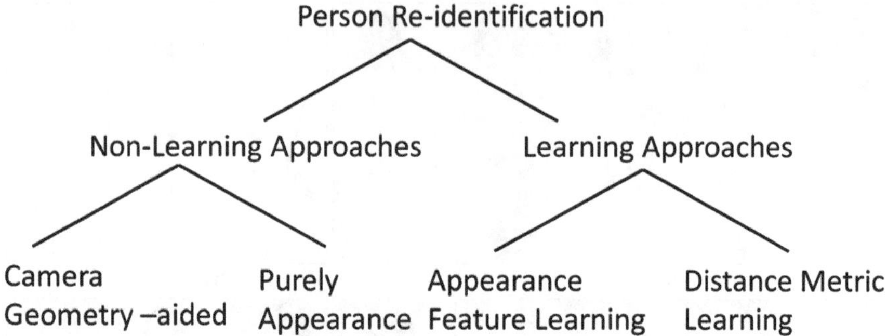

Fig. 5 A taxonomy of person re-identification approaches

3.1 Non-learning Approaches

In these approaches the re-identification model generation or model matching does not require a training step. Predefined appearance features are extracted from the video data and matched using fixed metrics. These methods can be further classified on the basis of whether or not camera calibration or homography estimation is required to complete the re-identification process. They can be classified as camera geometry-aided or purely appearance-based methods.

3.1.1 Camera Geometry-Aided Approaches

Homography is used for 3D location and person height estimation in [7], whereas principal axis based person matching is achieved in [8] using homographic transformations of the axes. Reliable homography estimation requires at least four corresponding points between camera views, but in wide area camera networks, camera FOVs are non-overlapping. A panoramic appearance map (PAM) proposed in [9] generates a single person signature extracted from multiple camera views of the person through camera calibration, triangulation, and 3D location estimation. However, the person needs to be visible in at least three cameras with overlapping views simultaneously, which can prove limiting is many cases. In [10] the placement and orientation of the 3D person model is determined using camera calibration and tracking information to aid the matching process. All of these methods use appearance features for computation of matching scores to achieve re-identification. Such methods impose the constraint that camera calibration for every camera in the network would be required.

3.1.2 Purely Appearance-Based Approaches

In these approaches the re-identification process is entirely based on appearance features. The goal is to extract discriminative and robust appearance models for re-identification.

In order to model the changes in appearance of objects between two cameras, a brightness transfer function (BTF) between each pair of cameras is learned from training data in [11, 12]. The BTF is used as a cue in establishing appearance correspondence. The human body is divided into multiple polar bins, and color and edge features extracted from each bin form the appearance descriptor in [13]. A match is established using correlation based similarity measure. Gheissari et al. [5] proposed a novel spatiotemporal segmentation algorithm based on watershed segmentation and graph partitioning to detect stable spatiotemporal edges called edgels. The appearance of the person is modeled as a combination of color and edgel histograms, and histogram intersection is used to establish a match between observations. Hamdoun et al. [14] utilize interest point based descriptors collected over a number of images of a person to characterize the person's appearance. The descriptors are matched using sum of absolute differences metric and re-identification is established using a best bin first (BBF) search on a KD-tree containing all gallery models. Spatial covariance regions are extracted from human body parts in [15] and spatial pyramid matching is used to design a dissimilarity measure. In an extension of this work [16], the covariance descriptors are combined to extract the mean Riemannian covariance matrices over time. These Riemannian covariance matrices are extracted from a dense grid to characterize the person's appearance.

The human silhouette is represented by two complementary appearance features: HSV histograms and recurrent local texture patches extracted over multiple images in [2]. In a further improvement to this model proposed in [4], the human silhouette is divided into head, torso and legs regions by detecting two horizontal axes of asymmetry and one vertical axis of symmetry. Each part is then described using three features; weighted HSV histogram, maximally stable color regions (MSCR), and recurrent highly textured local patches. The features extracted are combined over multiple images of a person to form an appearance model called Symmetry Driven Accumulation of Local Features (SDALF). Pictorial structures model was employed by Cheng et al. [17] for part based human detection and each part is used to extract HSV histograms and MSCRs. The part based representation is then used for re-identification. They propose a slight modification of pictorial structures to better localize body parts using multiple images of a person to guide the MAP estimates of the body configuration. This aids the extraction of more reliable visual features to improve re-identification.

3.2 Learning Approaches

These approaches are methods that require learning or training phase in order to build the appearance model or establish a match. These can be further classified into two categories depending on whether the learning is employed to learn optimal appearance features/build descriptors or to learn optimal distance metrics for re-identification.

3.2.1 Appearance Feature Learning

The human silhouette is represented as a collection of shape labels and the appearance descriptor is constructed using the spatial occurrence of the appearance labels with respect to each shape label in [2]. Since the appearance model is based on appearance labels learned on a training dataset, the generalizability of the model can be difficult. Similar appearance labels and visual context using local and global spatial relationships are used to describe an individual's appearance in [18]. Groups to which people belong to are used as the visual context to reduce ambiguity in person re-identification. Spatial distributions of self similarities with respect to appearance labels are learned and combined as the appearance descriptor in [3]. Re-identification is established between person images using a nearest neighbor classifier.

The re-identification model is generated from tracking data in [19]. The model is generated by encoding SIFT features extracted during tracking by Implicit Shape Model codebook learned offline. The spatial distance between the SIFT points also contributes towards the model. The high dimensional appearance model matching is computationally very expensive and the major drawback of this approach. Adaboost learning is employed by [20] to simultaneously learn discriminant features and ensemble of weak classifiers for pedestrian recognition. Weak classifiers are designed using a training set to determine the features that impart maximum discriminative ability. Partial least squares technique is employed to not only find discriminative weighting for color, texture and edge features, but also as a mechanism to reduce descriptor dimensionality in [21]. Haar like features and dominant color descriptors are used as features for re-identification and to guide detection of upper and lower body of a person in [22]. Adaboost classifier is used to find the most appropriate appearance model to use for matching images of people.

3.2.2 Distance Metric Learning

Methods belonging to this category shift the focus from building invariant appearance models to learning appropriate distance metrics that can maximize the matching accuracy regardless of the choice of appearance representation. A large margin nearest neighbor (LMNN-R) distance metric is learned in [23] such that it minimizes the distance between true matches and maximizes false match distances. The metric learned has the capacity to reject matches based on a universal threshold imposed on the matching cost. It was shown by Wang et al. [24] that by using a slight modification in the extraction of the feature vector stage, where overlapping regions of the human blob were used, the LMNN-R metric can provide greater robustness to re-identification under occlusion and scale changes. The person re-identification problem is treated as a relative ranking problem by Prosser et al. [25]. The main idea in their work was to learn a relative ranking of scores that capture the relevance of each likely match to the probe image, rather than comparing direct distance scores between correct and incorrect matches. In doing so, a set of weak SVM based rankers was learned using color and texture features on small training datasets, and combined using ensemble learning to realize a stronger ranker. A similar idea is explored by Zheng et al. [26]. They focus on maximizing the probability that a true match pair has a smaller distance than a false match pair.

3.3 Key Limitations

Existing methods developed to address the problem of person re-identification leverage clothing appearance based features. The key assumption across all methods is that people do not change clothes between occurrences. In general, there is a need to incorporate discriminative features that are not associated with clothing. Further, all of approaches discussed above address the single person re-identification problem. They do not present strategies to solve the multi-person re-identification problem in the context of wide area camera networks. Thus, developed models provide utility for closed set matching but may not scale for open set matching. In open set matching, the re-identification strategy has to explicitly or implicitly incorporate a false match rejection technique. This is a largely unaddressed aspect of person re-identification.

4 A Part-Based Spatio-Temporal Appearance Model

In the following, we discuss a strategy for multiple person re-identification with an implicit false match rejection criteria embedded in the framework [27]. A part-based person representation is used that combines color and facial features [28]. The model implicitly encodes the structural information as well as determines corresponding parts from the images to be matched. Body parts are described by color features that are used to build the person's appearance model, and used to establish a match. Features extracted from each body part are combined temporally into a model that describes the appearance along both the spatial and temporal

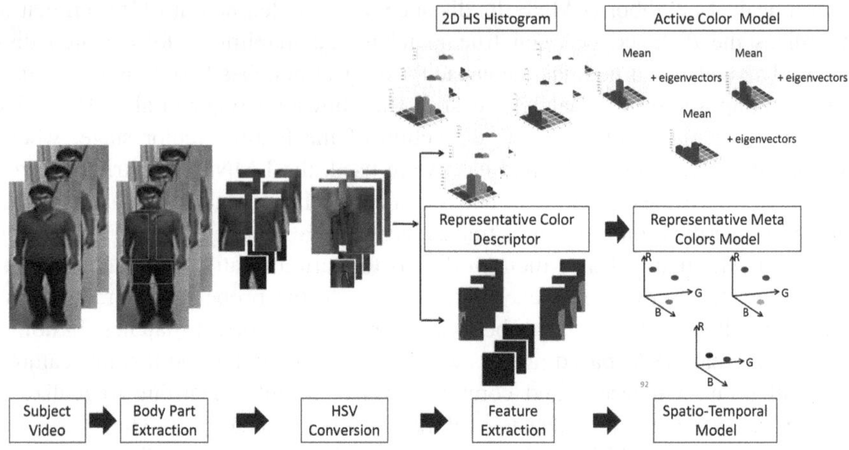

Fig. 6 Part-based color model generation pipeline

dimensions, making it a true spatio-temporal model. Further, depending on the presence of usable face images, the model can decide to include the facial features or exclude them, thus improving flexibility in an unconstrained environment. Finally, most of the current work tests the proposed appearance models in closed set single person re-identification scenarios while the presented model is evaluated in both open and closed set experiments for multiple person re-identification.

4.1 Spatio-Temporal Person Model Generation

The person model used in this approach is part-based and combines color and facial features to represent a person. Matching for re-identification has two steps: determining corresponding parts from the images to be matched and extracting appearance features from each corresponding part to establish a match. Different parts of the body can have different appearance features, for example shirt could be white and pants black. This observation has prompted us to adopt segmentation of the human body into meaningful parts before matching. This not only locates corresponding parts to be matched but also greatly reduces the probability of an incorrect match. Figure 6 shows an overview of the clothing color feature extraction and appearance model generation pipeline.

4.1.1 Body Part Extraction

Three stable parts are used represent the body of a person. This enables us to use a set of corresponding body parts for matching while imparting partial pose invariance to the overall model since individual body parts have fewer valid poses compared to the entire body. The parts extracted are not aligned with anatomical body parts and hence pose variations do not result in drastic changes in the detected parts. These body parts were extracted using the model proposed in [29], which models the human body as a collection of parts arranged in a deformable configuration. We use the six-part person model trained on the VOC 2008 pedestrian dataset [30]. Of the six parts, we retain only three stable parts: left torso, right torso and the upper legs, since they encapsulate the area of the body that provides maximum distinguishing appearance information. The model consists of a global root template for the entire body and local part filters. The global and local templates are based on Histogram of Oriented Gradients (HOG) features introduced in [31] and are learned during a training phase using latent SVM. The local part model consists of the spatial model, i.e. a set of allowable placements relative to the detection bounding box and the part filter.

4.1.2 Color Model Extraction

Color is the most expressive and powerful cue for object recognition and is leveraged in our model to characterize appearance. We extract two different color descriptors for each body part. The first descriptor is based on a color histogram that characterizes the distribution of colors within the body part and the second descriptor is a set of representative colors.

Active Color Model

The HSV color space is a perceptually intuitive color space, defining color in terms of its hue, purity or saturation, and brightness or value. Thus, it is a useful color space to match colors in, or determine if one color is similar to another consistent with human color perception. A 2D histogram based on the H and S color channels of the HSV color space is used to characterize the chromatic content of each body part. Hue is invariant to changes and shifts in the illumination intensity while saturation is not [32]. Each channel is quantized into five bins resulting in a 5×5 element HS histogram.

In order to capture the appearance variations of the chromatic content over time, we build a probabilistic model following the idea of Active Appearance Models [33] for each body part based on the underlying color histogram. Sequence of frames are used to extract the 2D color histograms. Thus, the sequence of 2D histograms forms the training set used to build the Active Color Model (ACM). The appearance model is given by,

$$g = \bar{g} + A_g b_g$$

and it captures variations in the 2D histograms across the sequence of frames. Here, g is the mean 2D histogram and A_g is the matrix describing the modes of

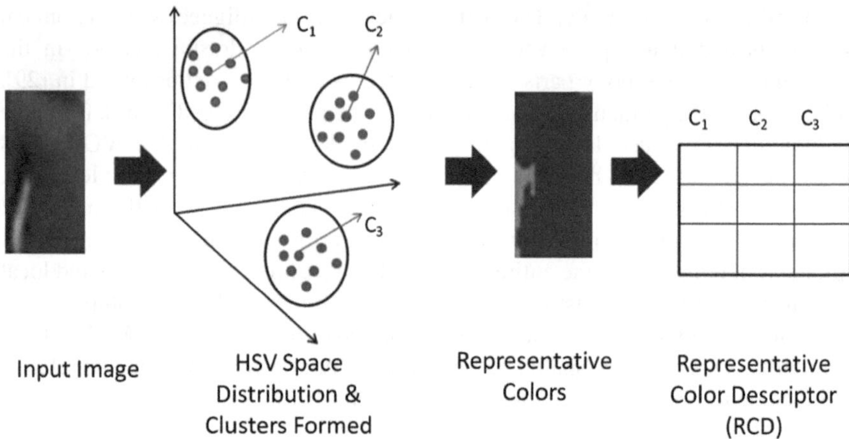

Fig. 7 Representative color descriptor generation process

variations in the color histograms within the sequence of frames. Vector b_g is the parameter set of the ACM. In our experiments, we only use columns of A_g that retain 75 % of the variations. This not only helps in capturing the maximum variations but also helps to eliminate redundant information and outliers.

Representative Meta Colors Model

Every part is characterized using a set of representative colors. These representative colors are extracted by fitting finite mixture models to the color vector using the method proposed in [34]. This method employs a minimum message length type selection criterion that is part of the Expectation–Maximization algorithm. This unsupervised selection and estimation of clusters is based on the Fisher information matrix. This clustering ensures that the representative colors are formed in such a way that the intra-cluster variance is minimized while the inter-cluster distance is maximized. Figure 7 is a pictorial representation of the clustering process that results in extraction of the representative colors.

Clustering is performed in the HSV space as it is perceptually intuitive. This maximizes our ability to capture the perceptually dominant color in the presence of illumination changes. RGB triplet is used to represent the final clusters. Each representative color cluster is described using the cluster's average color. The representative colors descriptor (RCD) is defined as,

$$RCD = (\{C_i\}|i = 1, \ldots, N_c)$$

where, N_c is the number of color clusters. In our implementation $N_c \leq 3$.

To meaningfully combine the representative color descriptors extracted from the sequence of frames, the representative colors are matched from frame to frame.

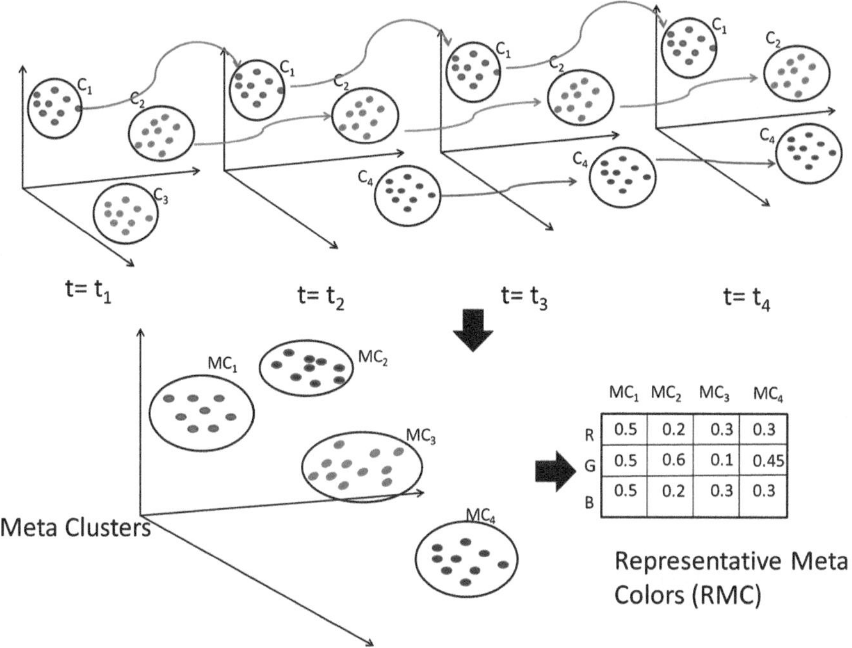

Fig. 8 Representative meta colors generation process

The frame-to-frame correspondence is established in the RGB space. Figure 8 illustrates the frame-to-frame representative colors matching and subsequent meta-colors determination. The representative colors that do not match for more than ten consecutive frames are rejected. Finally, another level of clustering is performed within the set of temporally matched colors. Once again, clustering is done in the HSV space but the corresponding RGB triplets are used to represent the meta-clusters. Within each set of matched colors, up to three representative meta-color clusters are built. Each representative meta-color cluster or Representative Meta Colors Model (RMC) is in turn described using the average color of the meta-cluster. These RMCs are extracted for each body part to form the representative color based spatio-temporal model. The RMC is defined as,

$$RMC = (\{MC_i\}|i = 1, \ldots, N_{mc})$$

where, N_{mc} is the number of meta color clusters formed over the sequence.

4.1.3 Facial Features Based Model

According to the study presented in [29], pixel gray level values of low resolution images as features can achieve a high face recognition rate. Such facial features are better suited for re-identification scenarios and we leverage this attribute. Since

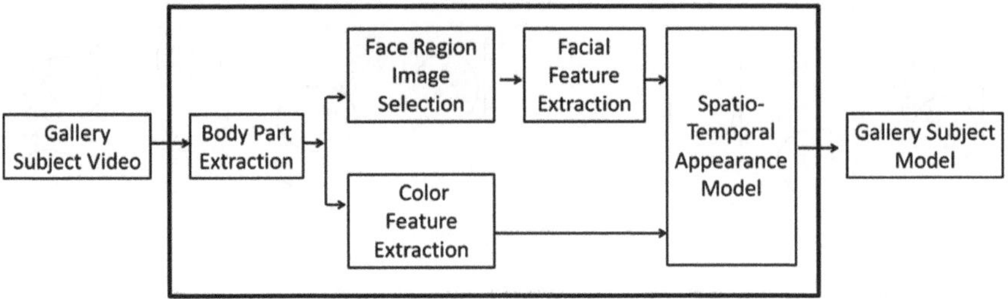

Fig. 9 Combined color and facial features model generation pipeline

the head region images could be blurred or no face could be present for a detected region, facial feature extraction is not always possible. Low-level image cues are used to select usable face images. If usable face images are present, the person's face model is built by using the facial features. As a result of the selection process, the face images are no longer temporally adjacent, even if they come from the same video. Thus, multiple models are learned for the same person's face to keep model inaccuracies due to feature misalignment to a minimum. Depending on the presence of usable face images, our method can decide to include the facial features or exclude them, this giving our model flexibility in an unconstrained environment.

Since the facial features are simply the image pixels, noisy or non-facial pixels could misguide the facial feature extraction severely. Hence a robust face image selection process is adopted before the facial features model is generated. Figure 9 depicts the pipeline from input video to output model generation with multiple feature integration.

Face Image Selection

The body part detector provides an estimate of the head region of a person and in order to extract facial features only the head regions with faces are retained. All the images are converted to gray scale and a 2-step selection process is used to retain usable face images. First, a threshold (τ_1) on RMS contrast is employed to reject incorrectly detected head region images as well as low contrast faces. The RMS contrast is computed as:

$$RMS\ Contrast = \sqrt{\frac{1}{wh}\left(\sum_{\mathrm{i}\in \mathrm{w}}\sum_{\mathrm{j}\in \mathrm{h}}(\mathrm{I_{ij}}-\bar{\mathrm{I}})^2\right)}$$

where, I_{ij} is the pixel intensity extracted from the face region image of size $w \times h$, and $\bar{I}$ is the mean intensity of the image pixels. In the next step, the retained images are then subjected to canny edge detection to detect prominent edges. An ellipse is fitted and if more than half the numbers of pixels in the ellipse fitted region are

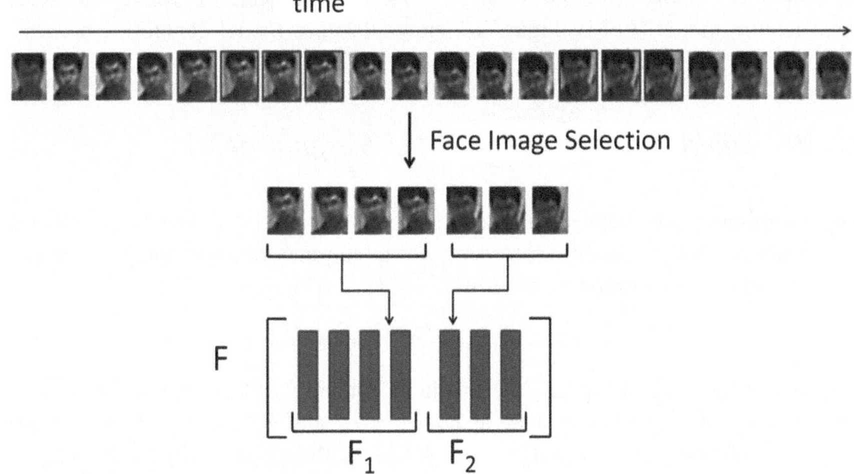

Fig. 10 Facial features model generation process

above a threshold (τ_2) then the head image is retained and used to extract the facial features. Consider n_{total} as the original head region images and the number of images retained after the selection process as n_{sel}. These images are not constrained to be temporally adjacent and if $n_{\text{sel}} < 2$, then the facial model cannot be generated. In our study, we use $\tau_1 = 0.04$ and $\tau_2 = 0.6$.

Facial Features Model Generation

The selected facial images are used to extract the facial features. All the head region images are resized to a fixed dimension of 24×20. The facial region images are vectorized by stacking its columns, and these vectors are used as the facial feature. Thus, the facial feature based model is simply a matrix $F = [v_{1,\ldots,}v_n]$, where v_k is the column corresponding to the *kth* face region. The matrix F will be of size $m \times n$, where n is the number of selected face images.

$$F = \begin{bmatrix} f_{1,1} & \cdots & f_{1,n} \\ \vdots & \ddots & \vdots \\ f_{m,1} & \cdots & f_{m,n} \end{bmatrix}$$

To minimize the errors in the model, the underlying features should be as aligned as possible. We make an assumption that among the selected face images, temporally adjacent frames are reasonably aligned with small changes in pose. Thus, within F, F_j is a submatrix of size $m \times n_j$, such that the columns of F_j are temporally consecutive face images. The submatrix is treated as a sub class, i.e. even if all columns from F belong to the same subject, the submatrices within F

are treated as visually different instances of the same subject. F is referred to as the facial feature model (FFM). Figure 10 shows the face model generation process.

4.2 Re-identification by Rectangular Assignment

The dissimilarity cost between a given gallery subject (G) and a probe subject (P) is a weighted sum of dissimilarity costs computed according to the color model and the facial features model, given as:

$$d(G,P) = w_{color} d_{color}(G,P) + w_{face} d_{face}(G,P).$$

If usable face region images are available for both gallery and probe subject, then the relevant weights are given as $w_{color} = 0.7$ and $w_{face} = 0.3$. If face region images are not available either for the probe or for the gallery subject, the weights used are $w_{color} = 1$ and $w_{face} = 0$, thereby utilizing only the color model.

Since the number of probe frames for an observed person can be greater than 1, the gallery subject model (G) is compared to N randomly selected frames of the probe subject (P) for matching. Color model cost is simply the minimum cost among all the probe frames. The color model cost is computed as:

$$d_{color}(G,P) = w_{ACM} d_{ACM}(G,P) + w_{RMC} d_{RMC}(G,P).$$

Both the color features contribute equally to the color model cost, hence $w_{ACM} = w_{RMC} = 0.5$. The ACM cost is the reconstruction error obtained by projecting the probe frame histogram into the ACM space of the gallery model and is given by

$$d_{ACM} = \|g_m - \hat{g}\|,$$

where $g_m = \bar{g} + A_g.\hat{b}$, $\hat{b} = A_g^{-1}(\hat{g} - \bar{g})$, and $\hat{g}$ denotes the probe histogram.

The RMC cost is based on the matching of the gallery RMC model to the RCD extracted from the probe frame. The gallery RMC and the probe RCD are both treated as signatures and the matching is treated as a transportation problem. The matching cost is calculated using the Earth Mover's distance (EMD) [35]. The color dissimilarity cost is computed separately for the two torso and legs parts. The overall body dissimilarity is computed as a Euclidean norm of cost obtained from each body part.

The facial feature matching is established using Sparse Representation in the context of face recognition [36]. The underlying implication is that given a dictionary matrix built using labeled training images of several subjects, a test image of the *ith* subject is the linear combination of the training images of only the same subject from the dictionary. In our case, a given gallery FFM contains all usable images of the gallery subject but the images are arranged into subsets depending on their temporal adjacency. Given a gallery FFM F_G, if the probe ID is same as the gallery ID, the probe image will lie approximately in a linear span defined by only a subset of the images that constitute the FFM. This implies that given a probe face image f_P, it can be expressed as $f_P = F_G \cdot \alpha$ and the objective is to find α that best generates f_P in F_G. Thus, among all possible solutions of α, we want the sparsest. This amounts to solving the following L1-minimization:

$$\hat{\alpha} = \arg\min \|\alpha\|_1 \text{ s.t.} f_P = F_G \alpha.$$

This optimization is solved using linear programming that leverages augmented Lagrange multipliers [37]. Hence, the facial features model cost is given by the following equation:

$$d_{face}(G, P) = \|f_P - F_G \hat{\alpha}\|_2,$$

and is an estimate of how well α reproduces f_P.

Re-identification is established between a pair of cameras, which means that there are multiple probes that are to be assigned to the appropriate gallery IDs simultaneously. Thus, model dissimilarity cost is computed for every pair of gallery and probe subject and a cost matrix is populated and used for overall assignment.

The cost matrix is given as $C(i,j) = d(G_i, P_j)$, where $i = 1, \cdots, N_G$ and $j = 1, \cdots, N_P$. N_P and N_G are the number of probe and gallery subjects, respectively. Since all the probe subjects come from the same video, their IDs are distinct so a one-to-one assignment between the probe and gallery sets is needed for multiple person re-identification. The combinatorial optimization is solved using the Munkres algorithm [38]. Given a cost matrix C, the one-to-one assignment is solved by minimizing the following objective function:

$$\sum_{i \in N_G} \sum_{j \in N_P} C(i,j) x_{ij},$$

where, x_{ij} represents assignment of subject G_i of the gallery set to subject P_j of the probe set, taking value 1 if assignment is done and 0 otherwise.

5 Evaluation

5.1 Dataset

There are a few public datasets that have been used to test re-identification models like ViPER [39], i-LIDS for re-identification [18], ETHZ [40] and CAVIAR4REID [17]. The ViPER dataset consists of images of people from two different camera views, but it has only one image of each person and each camera. ETHZ and i-LIDS has multiple images of people but they are all from the same camera. CAVIAR4REID has multiple images of a person but only from two different camera views. None of these datasets truly represent the videos obtained

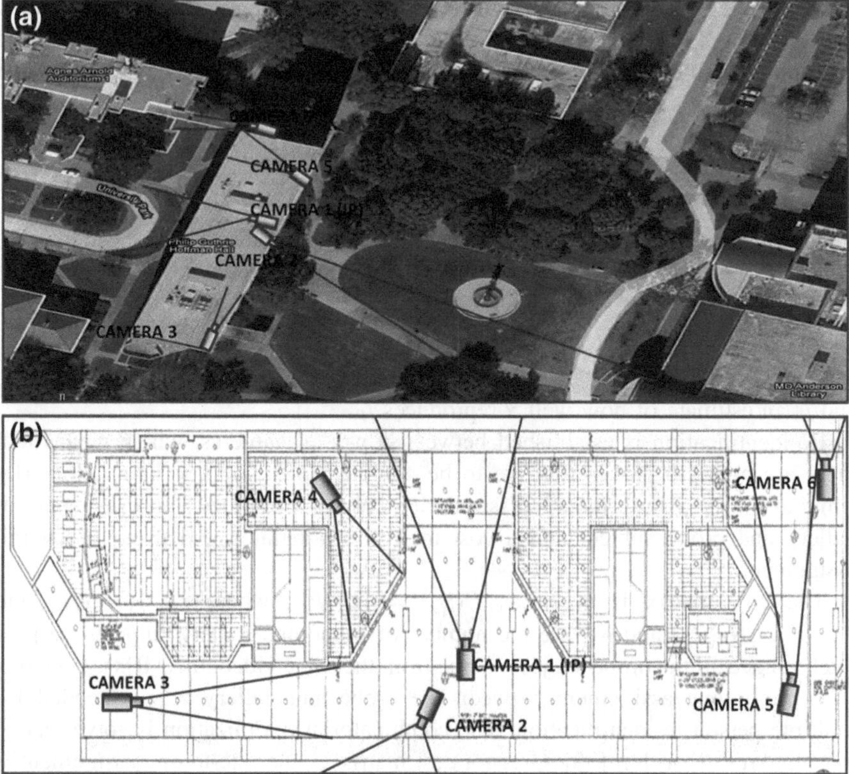

Fig. 11 Placement of outdoor cameras to monitor people entering or leaving a building

from a wide area camera network and are not optimal for evaluating multiple re-identification models. An appropriate dataset would consist of multiple images of a person and multiple camera views (more than 2). In addition, the images should capture variations with respect to pose and viewpoint changes, resolution changes, illumination conditions, and occlusions.

In order to capture a more realistic environment, we setup a camera network consisting of ten cameras in and around a building. The cameras were placed on the first floor and fifth floor of a building. Figure 11 shows the placement of cameras in the outdoor environment around the perimeter of the building wherein the camera positions are overlaid on a top view of the building as well as the floor plan of the building's first floor. Figure 12 shows example images from each of the outdoor cameras. Figures 13, 14 show the position and camera views of the indoor cameras, respectively. The re-identification data consists of 30 videos collected using nine of these ten cameras and consists of 40 subjects out of which 19 are used to establish true re-identification. The data is split into three scenarios based on the difference in environments between the camera pairs in which re-identification is to be established. The scenarios are Outdoor–Outdoor, Indoor–Indoor and

Fig. 12 Example frames obtained from outdoor cameras

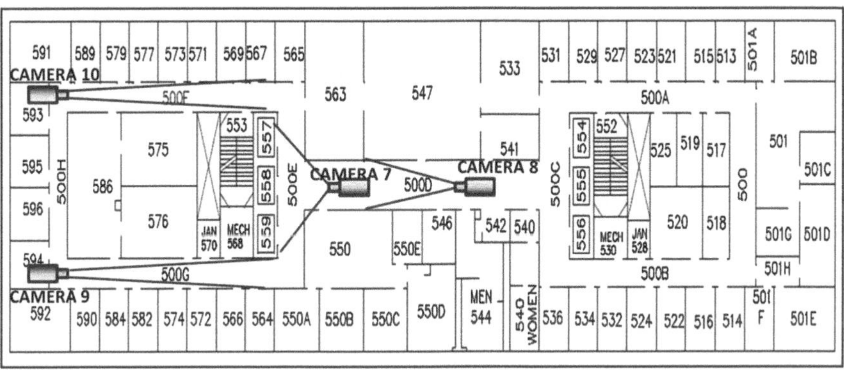

Fig. 13 Placement of indoor cameras on the 5th floor of a building

Outdoor–Indoor. In the Outdoor–Outdoor scenario there exists a large variation in the illumination conditions between cameras. The Indoor–Indoor scenario presents a more uniform illumination variation but is useful to test the model's discriminant capability. The Outdoor-Indoor scenario needs a good balance of discriminative capability and photometric invariance in the model in order to establish correct matches.

Within each category, re-identification is established with one or two spatial neighbors of a given camera. The spatial relationships are used only for the purposes of experimental setup and are not required to establish the re-identification. No prior information about the spatio-temporal relationships is leveraged in this approach. The acquired data is manually annotated. This not only provides the

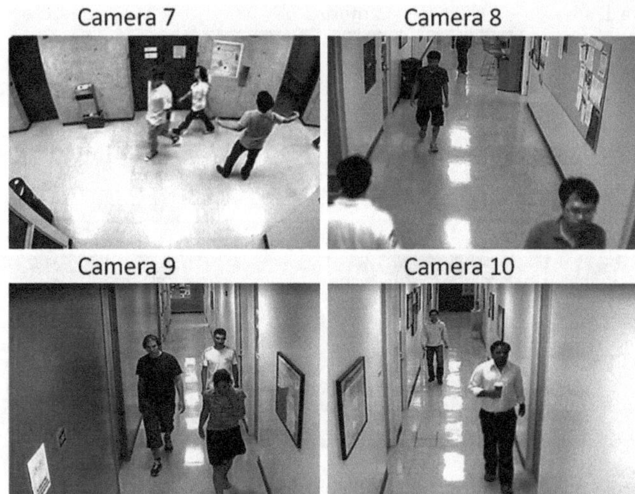

Fig. 14 Example images obtained from indoor cameras

tracking data needed for re-identification but also provides ground truth for evaluation of the results. The person bounding boxes obtained from manual annotations are all normalized to 128×64.

For a given camera pair, the gallery model for each subject is built using all the available gallery frames. Only N randomly selected frames from a probe sequence are used to calculate the dissimilarity cost. In all experiments performed, we chose $N = 10$ and results presented are based on analysis of 50 independent trials.

5.2 Experiments

To evaluate our model under a multiple person re-identification framework we designed two kinds of experiments.

Closed Set Experiments: Only a subset of the probe set, i.e. an intersection between the probe and gallery set is used to establish re-identification. This implies that subjects in the closed probe set are in the intersection between the probe IDs and gallery IDs. This is a closed set experiment, in the sense that it is not possible to have false positive but rather only correct matches or mismatches. False positive is defined as the model matching a probe ID incorrectly to a gallery ID when in fact the probe ID is not in the gallery set. This experiment is more geared towards testing the sensitivity of the appearance model.

Open Set Experiments: The entire probe set is used for re-identification wherein all the subjects in the probe set might not be present in the gallery set. The probe and the gallery are truly open sets implying the possibility of false positives. This experiment is designed to test the appearance based false acceptance reduction

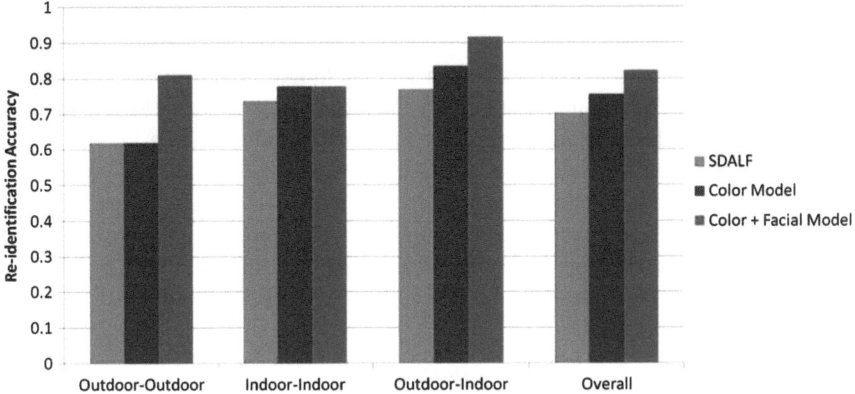

Fig. 15 Closed set re-identification results

criteria and a secondary objective of testing the generality of the model. In order to reduce the false acceptance, a threshold is imposed on the cost matrix C prior to multiple person re-identification.

5.2.1 Closed Set Experimental Results

The closed set results are evaluated using matching accuracy, i.e. number of probe subjects matched correctly. In other words, it is the percentage of rank-1 correct matches. The effectiveness of the presented model is compared to the SDALF model proposed in [4].

Figure 15 shows the closed set re-identification performance using two variants of the model, one based only on color features and the other using both the color and facial features. As seen, the value of incorporating facial features into the spatio-temporal model is clearly evident. In both the Outdoor–Outdoor and Outdoor–Indoor scenarios, we observe a significant improvement in the re-identification accuracy. In the Indoor–Indoor scenario the performance remains unchanged by addition of the facial features but also does not have an adverse effect on the re-identification rate. This implies that even low-resolution facial regions with varying illumination and pose can contribute towards improving the discriminative ability of our model. In the Outdoor–Outdoor scenario, adding facial features causes the accuracy to increase from 62–80 %, which is a considerable improvement. The overall accuracy increases to 83 % from 75 %. Both variants of the model outperform the SDALF model in all three scenarios. In all the three scenarios, color only model gives 75 % accuracy compared to 70 % using SDALF.

5.2.2 Open Set Experimental Results

In the case of re-identification under open set matching, true positives (TPs) are the number of probe IDs that are correctly matched, mismatches (MMs) are the number of probe IDs that are incorrectly matched to gallery when that probe ID does exist in the gallery, and false positives (FPs) are the number of probes IDs that are matched to the gallery when the probe ID does not exist in the gallery.

In order to reduce false acceptances, a threshold is imposed on the cost matrix C prior to assignment. The open set results are presented in terms of Accuracy vs. false acceptance rate (FAR) curves. The accuracy and FAR are defined as:

$$Accuracy = \frac{TPs}{N_P}$$

$$FAR = \frac{MMs + FPs}{N_P},$$

where, N_P denotes the total number of probe subjects. The curves are obtained by varying a threshold imposed on the matching cost during the computation of the cost matrix. The threshold is varied from 0 to 0.9 in increments of 0.05. Two different assignment techniques are employed: optimal (Munkres) and the sub-optimal assignment. The sub-optimal assignment technique is usually used when the cost matrices have many forbidden assignments. The assignment is suboptimal in the sense that the overall assignment cost is not the minimum possible value. In case of re-identification, the gallery is ever increasing and most likely the intersection between gallery and probe is small compared to the size of the gallery. Thus, the possibility of incorrect assignments increases and suboptimal assignment technique could be better suited for such cases.

Figure 16 shows that in all three scenarios it is possible to find a threshold that yields the best possible trade off between accuracy and FAR. In the Outdoor–Outdoor scenario a threshold of 0.5 yields the best possible Accuracy/FAR ratio of 60 %/33 % using the color and facial features model and optimal assignment technique. In the Indoor–Indoor and Outdoor–Indoor scenarios as well, color and facial features model gets the best Accuracy/FAR ratio possible. The exact same observations are applicable to the suboptimal assignment technique. Except in the case of the Outdoor–Indoor scenario under suboptimal assignment, our color only model is able to find a threshold that gives better Accuracy/FAR ratio compared to SDALF. In the exception case, SDALF finds a threshold that gives nominally better Accuracy/FAR ratio compared to the color only model. Thus, overall in both variants of the model, it is possible to find a threshold that gives better Accuracy/FAR ratio than SDALF. This implies that this new model can generalize well to handle open set re-identification.

In order to compare the open set performance of the three models (SDALF, color only, and color and facial features), we use the discrimination measure as an indicator of model's accuracy and false rejection ability. The discrimination

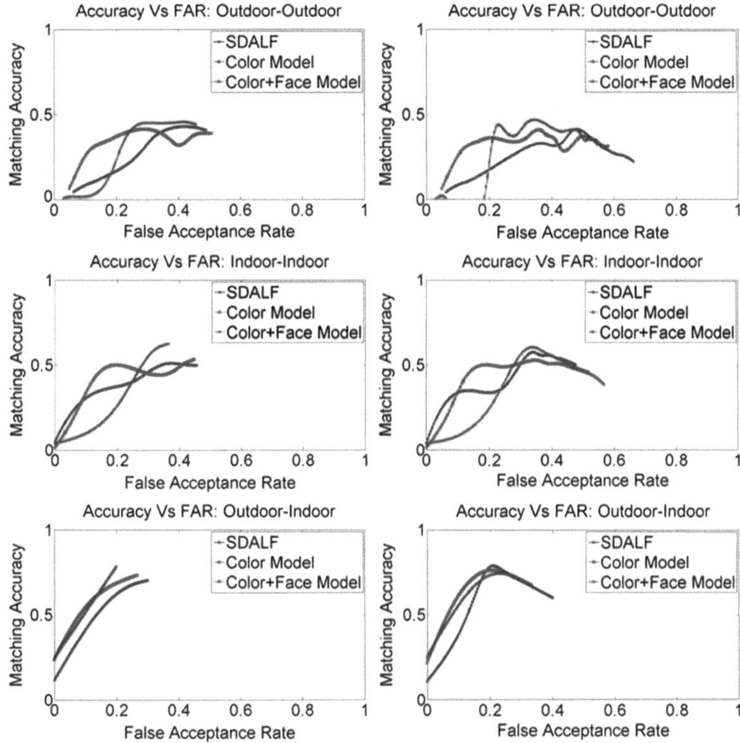

Fig. 16 Open set matching accuracy vs. FAR results: **a** optimal assignment, and **b** suboptimal assignment. The *top*, *middle* and *bottom* row are results obtained for the Outdoor–Outdoor, Indoor–Indoor and Outdoor–Indoor scenarios, respectively

Table 1 Open set (discrimination) results for all the models

Assignment	Method	Outdoor–outdoor	Outdoor-Indoor	Indoor–Indoor
Optimal (Munkres)	SDALF	0.38	0.35	0.32
	Color model	0.34	0.35	0.32
	Color + facial model	0.38	0.4	0.4
Sub-optimal	SDALF	0.34	0.3	0.28
	Color model	0.33	0.26	0.29
	Color + facial model	0.37	0.32	0.30

measure is simply the area between the curve and the non-discrimination line. Table 1 shows that in all three scenarios the color and facial features model has improved discrimination over the other two models evaluated. In the Outdoor–Outdoor scenario and Indoor–Indoor scenario using optimal assignment technique, and Indoor–Indoor scenario using suboptimal assignment technique, SDALF has a higher discrimination than our color only model. This can be attributed to the fact that SDALF includes a texture component in its model that is absent in ours. Overall, all the models suffer a slight drop in discrimination using the suboptimal assignment technique.

In general, closed and open set experiments both suggest that using color and facial features has a distinct advantage for re-identification over using only color or even combined color and texture features (SDALF). From the shape of these curves we can deduce that a global threshold on the matching cost is a crude yet sensible strategy for minimizing the false acceptance rate. As the threshold increases the accuracy values start increasing as well with a less drastic increase in FAR. The most important conclusion that we can draw from these experiments is that the proposed model can be used for false acceptance reduction during re-identification. Specifically, open set experiments can be used to select the parameters for false acceptance reduction criteria, which in our case is a suitable threshold.

6 Discussion

In this chapter, we have introduced the person re-identification problem in the context of wide area camera networks and its challenges. A strategy for multiple person re-identification with an implicit false match rejection technique is discussed. We have presented a spatio-temporal model based on color and facial features that captures complementary aspects of a person's appearance. A face model based on useful facial features is exploited using sparse representation and applied towards a multi-feature model. A low-level image feature based face region selection module is designed to select usable face region images.

A wide area camera network is deployed to acquire representative data from multiple cameras, spanning outdoor and indoor environments. Open and closed set experiments were performed on data collected from nine cameras and results presented indicate that the developed model has the necessary balance of discriminative ability and photometric invariance. Appropriate open and closed set performance metrics were also discussed.

We would also like to point out that, to the best of our knowledge, this is the first study that offers strategies for principled multi-parametric spatio- temporal appearance model generation, multiple person re-identification, and appearance model based criterion for reduction in false acceptance rates. Future work in addressing this challenging problem should focus on enhancing any model's discriminative and invariant properties. Overall, exploring better solutions to the multiple person re-identification and false acceptance reduction will lead to improvements in overall performance for a multitude of applications that would utilize wide area camera networks.

References

1. Sivic, J. Zitnick, C. L., Szeliski, R.: Finding people in repeated shots of the same scene. In: British Machine Vision Conference, pp. 909–918 (2006)
2. Bazani, L., et al.: Multiple-shot person re-identification by HPE signature. In: 20th International Conference on Pattern Recognition (ICPR), pp. 1413–1416 (2010)
3. Cai, Y., Pietikäinen, M: Person re-identification based on global color context. In: Proceedings of the 2010 International Conference on Computer Vision, Queenstown, New Zealand (2011)
4. Farenzena, M., et al.: Person re-identification by symmetry-driven accumulation of local features. In: IEEE Conference on Computer Vision and Pattern Recognition (CVPR), pp. 2360–2367 (2010)
5. Gheissari, N., Sebastian, T.B., Hartley, R.: Person reidentification using spatiotemporal appearance. In: IEEE Computer Society Conference on Computer Vision and Pattern Recognition, pp. 1528–1535 (2006)
6. Xiaogang, W., et al.: Shape and appearance context modeling. In: IEEE 11th International Conference on Computer Vision (ICCV), pp. 1–8 (2007)
7. Park, U., et al.: ViSE: visual search engine using multiple networked cameras. In: 18th International Conference on Pattern Recognition (ICPR), pp. 1204–1207 (2006)
8. Hu, W. et al.: Principal axis-based correspondence between multiple cameras for people tracking. In: IEEE Transactions on Pattern Analysis and Machine Intelligence, vol. 28, pp. 663–671 (2006)
9. Gandhi, T., Trivedi, M.M.: Person tracking and reidentification: introducing panoramic appearance map (PAM) for feature representation. Mach. Vis. Appl. **18**:207–220 (2007)
10. Baltieri, D., et al.: Multi-view people surveillance using 3D information. In: IEEE International Conference on Computer Vision Workshops (ICCV Workshops), pp. 1817–1824 (2011)
11. Porikli, F.: Inter-camera color calibration by correlation model function. In: Proceedings International Conference on Image Processing (ICIP), vol. 3, pp. 133–136 (2003)
12. Javed, O., Shafique, K., Rasheed, Z., Shah, M.: Modeling inter-camera space-time and appearance relationships for tracking across non-overlapping views. Comput. Vis. Image Underst. **109**:146–162 (2008)
13. Jinman, K., Cohen, I., Medioni, G.: Object reacquisition using invariant appearance model. In: Proceedings of the 17th International Conference on Pattern Recognition (ICPR), vol. 4, pp. 759–762 (2004)
14. Hamdoun, O., Moutarde, F., Stanciulescu, B., Steux, B.: Person re-identification in multi-camera system by signature based on interest point descriptors collected on short video sequences. In: Second ACM/IEEE International Conference on Distributed Smart Cameras (ICDSC), pp. 1–6 (2008)
15. Bąk, S., Corvée, E., Brémond, F., Thonnat, M.: Person re-identification using spatial covariance regions of human body parts. In: Seventh IEEE International Conference on Advanced Video and Signal Based Surveillance (AVSS), pp. 435–440 (2010)
16. Bąk, S., Corvée, E., Brémond, F., Thonnat, M.: Boosted human re-identification using Riemannian manifolds. Image Vis. Comput. 30(6–7) 443–452 (2012)
17. Cheng, D.S., Cristani, M., Stoppa, M., Bazzani, L., Murino, V.: Custom pictorial structures for re-identification. In: British Machine Vision Conference, pp. 68.1–68.11 (2011)
18. Zheng, W.-S., Gong, S., Xiang, T.: Associating groups of people. In: British Machine Vision Conference, London, UK (2009)
19. Kai, J., Bodensteiner, C., Arens, M.: Person re-identification in multi-camera networks. In: IEEE Computer Society Conference on Computer Vision and Pattern Recognition Workshops (CVPRW), pp. 55–61 (2011)
20. Gray, D., Tao, H.: Viewpoint invariant pedestrian recognition with an ensemble of localized features. In: European Conference on Computer Vision, Marseille, France (2008)

21. Schwartz, W.R., Davis, L.S.: Learning discriminative appearance-based models using partial least squares. In: XXII Brazilian Symposium on Computer Graphics and Image Processing (2009)
22. Bąk, S., Corvée, E., Brémond, F., Thonnat, M.: Person re-identification using Haar-based and DCD-based signature. In: Seventh IEEE International Conference on Advanced Video and Signal Based Surveillance (AVSS), pp. 1–8 (2010)
23. Dikmen, M., Akbas, E., Huang, T.S., Ahuja, N.: Pedestrian recognition with a learned metric. In: Asian Conference on Computer Vision, Queenstown, New Zealand (2011)
24. Wang, s., Lewandowski, M., Annesley, J., Orwell, J.: Re-identification of pedestrians with variable occlusion and scale. In: IEEE International Conference on Computer Vision Workshops (ICCV Workshops), pp. 1876–1882 (2011)
25. Prosser, B., Zheng, W.-S., Gong, S., Xiang, T.: Person re-identification by support vector ranking. In: British Machine Vision Conference (2010)
26. Zheng, W.-S., Gong, S., Xiang, T.: Person re-identification by probabilistic relative distance comparison. In: IEEE Conference on Computer Vision and Pattern Recognition (CVPR), pp. 649–656 (2011)
27. Bedagkar-Gala, A., Shah, S.K.: Multiple person re-identification using part based spatio-temporal color appearance model. In: IEEE International Conference on Computer Vision Workshops (ICCV Workshops), pp. 1721–1728 (2011)
28. Bedagkar-Gala, A., Shah, S.K.: Part-based spatio-temporal model for multi-person re-identification. Pattern Recogn. Lett. (2011)
29. Yang, A.Y., Wright, J., Ma, Y., Sastry, S.S.: Feature selection in face recognition: a sparse representation perspective. University of Illinois (2007)
30. Everingham, M., Van $\sim$ Gool, L., Williams, C.K.I., Winn, J., Zisserman, A.: The pascal visual object classes (VOC) challenge. Int. J. Comput. Vis. **88**:303–338 (2010)
31. Dalal, N., Triggs, B.: Histograms of oriented gradients for human detection. In: IEEE Computer Society Conference on Computer Vision and Pattern Recognition (2005)
32. van de Sande, K.E.A., Gevers, T., Snoek, C.G.M.: Evaluating color descriptors for object and scene recognition. IEEE Trans. Pattern. Anal. Mach. Intell. **32**:1582–1596 (2010)
33. Cootes, T.F., Edwards, G.J., Taylor, C.J.: Active appearance models. IEEE Trans. Pattern Anal. Mach. Intell. **23**:681–685 (2001)
34. Figueiredo, M.A.T., Jain, A. K.: Unsupervised learning of finite mixture models. IEEE Trans. Pattern. Analy. Mach. Intell. **24**:381–396 (2002)
35. Rubner, Y., Tomasi, C., Guibas, L.J.: The earth mover's distance as a metric for image retrieval. Int. J. Comput. **40**:99–121 (2000)
36. Wright, J., Ganesh, A., Zihan, Z., Wagner, A., Yi, M.: Robust face recognition via sparse representation. In: 8th IEEE International Conference on Automatic Face & Gesture Recognition FG, pp. 1–2 (2008)
37. Koko, J., Jehan-Besson, P: An augmented Lagrangian method for $\mathrm{TVg} + \mathrm{L1}$-norm minimization. J. Math. Imaging Vis. **38**:182–196 (2010)
38. Munkres, J.: Algorithms for the assignment and transportation problems. J. Soc. Ind. Appl. Math. **5**:32–38 (1957)
39. Gray, D., Brennan, S., Tao, H.: Evaluating appearance models for recognition, reacquisition, and tracking. In: 10th IEEE International Workshop on Performance Evaluation of Tracking and Surveillance (PETS) (2007)
40. Ess, A., Leibe, B., Van Gool, L.: Depth and appearance for mobile–scene analysis. In: IEEE 11th International Conference on Computer Vision (ICCV), pp. 1-8, (2007)

Opportunities and Challenges of Terrain Aided Navigation Systems for Aerial Surveillance by Unmanned Aerial Vehicles

Samil Temel and Numan Unaldi

Abstract Unmanned Aerial Vehicle (UAV) technology has become a promising means both for military and civilian surveillance issues that UAVs may provide more accurate, inexpensive and durable information than ground surveillance systems. The information can be obtained from various sensors which are equipped on a UAV. Most of the current UAVs depend on satellite based navigation systems such as Global Positioning System (GPS). However, GPS signals are easily jammed especially in military fields which necessitate a Terrain Aided Navigation (TAN) system. TAN systems aim to provide position estimates relative to known terrains. Such systems collect the height values from the surface with the help of active range sensors which are then matched within a terrain Digital Elevation Map (DEM). In this chapter, we have developed a preliminary TAN system as a testbed, in order to emphasize and address the opportunities and challenges of designing an autonomous navigation system. In addition, we have determined and summarized some of the design objectives for UAV based surveillance posts.

Keywords UAV surveillance · Terrain aided navigation · Simulated annealing

---

S. Temel · N. Unaldi (✉)
Turkish Air Force Academy (TUAFA), Istanbul, Turkey
e-mail: n.unaldi@hho.edu.tr

S. Temel
e-mail: s.temel@hho.edu.tr

Augment Vis Real (2014) 6: 163–177

DOI: 10.1007/8612_2012_6

Published Online: 12 September 2012

1 Introduction

A UAV is defined as an aerial vehicle that does not carry a human, uses aerodynamic force to provide lift, can fly autonomously or be piloted remotely and may carry lethal or nonlethal payloads [1]. Thanks to their versatility, flexibility, easily installation and relatively small expenses, usage of UAVs promise new ways for both military and civilian reconnaissance and surveillance applications. For the last two decades, in military areas, UAVs are being used for real-time surveillance, reconnaissance, intelligence and warfare operations. On the other hand, for civilian areas, UAVs are well suited for situations that are too harsh or dangerous for direct human monitoring.

One crucial advantage of UAVs is that they assist surveillance by improving coverage throughout the remote and unreachable sections of terrains. Hence UAVs can be defined as "eyes on the sky". As an example, for the last decade, UAVs are being used for Homeland Border Security [2]. Equipped with various electro-optical sensors (such as cameras) UAVs provide precise and real-time imagery to the border security ground control operators, who then successfully deploy border patrol agents to the exact location of the border. While the range of UAVs is wider compared to stationary surveillance equipment, UAVs would have a greater chance of tracking a border violator with thermal detection sensors than the stationary video equipment.

UAVs can also provide precious assessments after catastrophic natural events such as earthquakes, tsunamis, floods, hurricanes etc. As the ground surveillance systems can be destroyed or limited, as it happened during Japan earthquake in 2011, UAVs may provide autonomous, accurate and robust information for the search and rescue efforts. This information can vary from photos, video or radar images which will help to find survivors. If dispatching of ground rescue teams into the disaster area to examine the damage and find survivors is extremely dangerous or impossible, UAVs can play a crucial role for saving lives. For example, one significant rescue service is developed and successfully operated in UAVTech lab which presents an emergency UAV mission scenario [3]. It involves search and rescue for injured people by utilizing UAVs. The mission study is divided into two steps. In the first step; UAVs scan designated areas and try to identify injured civilians. In the second step, an attempt is made to deliver medical and other supplies to identified victims.

UAVs are also used to examine environmental and scientific facts. Surveillance of animal swarms, vegetation, volcanic mountains, forests etc. are such examples. As an example of a scientific study, which is totally based on deploying UAVs, is developed in Australian Center for Field Robotics (ACFR). In AFCR, a UAV based surveillance system is used to detect aquatic weeds in inaccessible habitats of Australia. Fixed-wing and rotary-wing UAV pairs are successfully deployed for large scale mapping and precision classification of woody/aquatic weed infestations [4].

Another example of UAV surveillance can be seen in forensic examinations where forensic events vary from car accidents to public security situations etc. Although there are many other sources of intelligence that provide information from a scene with manned helicopters, they cannot fly in close proximity to the sites like UAVs can do. Also in order to safely monitor a situation before personnel are sent into a dangerous environment, UAVs can provide lifesaving information to the authorities. An example of UAV usage in forensic scenes can be the DraganFly Heli-UAV which provides high definition motion video for security, reconnaissance, inspection, damage assessment, research, real estate promotion, or advertising etc. [5]. Equipped with IEEE 802.11n based radio equipment, DraganFly UAVs have been successfully used in various surveillance scenarios. Another example is the FIU-301 Unmanned Aerial System (UAS) of the Ontario Provincial Police department of North America [6]. The system was developed to obtain high quality digital aerial images of major case scenes in a timely and efficient manner while operating within a secure police environment.

Different versions of UAVs are also being developed to be used for the upcoming lunar and planetary posts [7, 8]. As, the world has recently turned its attention back to the moon, it has recently turned out to be a competition to land humans to the lunar surface [9]. Thus, various techniques must be developed or reinvented to place humans and cargo safely and precisely. Because there is no satellite navigation (such as GPS) data available on lunar surface, TAN systems can be extended to assist a lunar landing spacecraft with precise and safe landing. The required landing precision for the upcoming lunar missions is much stricter than Apollo 11 where the landing ellipsoid was 18.8 km along-track by 4.8 km across-track [7]. Although Apollo missions were able to achieve reasonably precise landings, the accuracy is not sufficient to meet the new objective, particularly for the unmanned missions that will not have benefit of astronaut-assisted navigation. To improve navigation precision, TAN is required, since inertial sensors alone cannot achieve the necessary performance.

It can be seen in the aforementioned examples that UAVs have a crucial potential to provide valuable surveillance assessments both for civilian, scientific and military applications. However, the autonomous and unmanned nature of the UAV necessitates accurate navigation capabilities. In aeronautics, consistent, incessant and exact location information is vitally important. Hence, most of the UAV systems are equipped with satellite navigation systems such as GPS, GLONASS and GALILEO. These systems are widely used with Inertial Navigation Systems (INS) in order to provide an air vehicle with continuous navigation information [10, 11]. Although satellite navigation systems provide about 10 m location accuracy, they are highly vulnerable to jamming and cannot be used in lunar and underwater missions [12]. In addition, if a navigation instrument such as a GPS receiver defects, an auxiliary system has to be assigned to estimate the location of a UAV. Thereby, there has always been a need for a satellite-navigation-free TAN system for UAVs.

Development of TAN systems started in 1970s and they were successfully used for Tomahawk cruise missile navigation [13]. Nowadays, TAN technology has

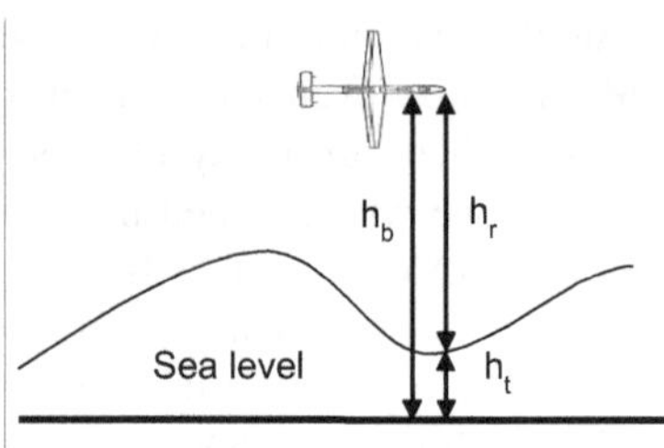

Fig. 1 Calculation of the height of the terrain from a UAV

reached to be used in underwater, planetary or ground vehicles [14]. However, it hasn't been successfully adapted to any UAV systems yet.

The basic idea behind a TAN system is to match the height values collected from a radar altimeter with the terrain heightmap data of an area of Interest (AOI) which is loaded into the flight computer in advance of a flight. While a UAV flies along its journey, the height of the terrain below the air vehicle is estimated with the help of a radio altimeter and a barometric altimeter as shown in Fig. 1. The height of the terrain h_t is calculated by subtracting the radar altimeter height h_r from barometric altimeter height h_b which is shown in (1). Measurements are taken periodically and when sufficient height values are collected, they are searched within the AOI which the UAV is currently flying over.

$$h_t = h_b - h_r \tag{1}$$

Although TAN systems provide various advantages for UAV navigation, there are some inherited disadvantages. Four of the major ones can be summarized as follows:

(i) TAN systems need undulating or rough terrains to operate [9],
(ii) The search and solution spaces of a TAN system is so huge that with a deterministic algorithm, providing a UAV with exact location information in a feasible amount of time is nearly impossible [8],
(iii) The active range sensors produce errors on high altitudes,
(iv) The miss probability of the matching the height values (that a UAV takes along its journey) with the DEM data is high because of the resolution gaps on the DEM data.

These problems necessitate the usage of a metaheuristic TAN approach for determining the location of a UAV. In this chapter, we address the issue of autonomous navigation, that is, the ability for a navigation system to provide information about the states of a vehicle without the need for a priori infrastructure such as GPS data. In this study, we have developed an exemplary TAN system in order to intensify and address the opportunities and challenges of designing an autonomous navigation system. The consequences of our research revealed that metaheuristic algorithms (such as SA) can be a good alternative for the determination of a vehicle's location. Thus, we have developed a TAN system which is based on SA approach and conducted our work on a real 3D DEM heightmap of northern part of Turkey which is nearly 100 × 100 km wide with 30 m resolution.

We expanded our work both on respectively flat and rough zones on the terrain. Our studies revealed that although average position estimation error rates on flat terrains are higher compared to rough terrains, a UAV can autonomously determine its location with mitigated drifts even on relatively flat zones. To our knowledge this is the first study which explores the TAN design objectives and with metaheuristics. We hope that our study will reveal and enlighten many of the controversial issues in the literature.

The related work, overview of TAN systems, SA algorithm, our exemplary TAN method and the conclusion with design objectives are described in following subsections.

2 Related Work

The term "*navigation*" originates from the Latin word "*navigare*" which means "*sailing ships*". For more than a century, the term is also used in aviation terminology to indicate determining the geographical position and velocity of air vehicles. In early days of aviation, pilots used to determine their location by matching particular land forms with the topographical printed maps while flying. Today, the range of airborne navigation systems and their capabilities are greater now than at any other time in aviation history.

Mainly, airborne navigation systems are divided into two main categories: *INS navigation* and *reference based navigation* [8]. An INS system utilizes an inertial measurement (which is basically a compass) device to determine an air vehicle's position, velocity, and attitude at high data rates. These data has a vital role for the guidance and control of an air vehicle. However, the INS equipment continually produces position estimation errors which necessitate *absolute sensors* to constrain the drift. Absolute sensors are categorized into two groups: *satellite based* and *terrain based*. Generally, satellite based navigation systems utilize GPS sensors. Fusion of INS with GPS systems has been widely studied in the literature [15, 16]. On the other hand, terrain based systems diminish the need for GPS devices with deploying a radar (or sometimes laser) altimeter and loading the terrain height map database on to an air vehicle. Although satellite navigation systems are well suited to UAV navigation, if the mission is within a GPS denied environment, such as a military zone where the signal is interfered, for underwater or lunar systems where no GPS signal is available, etc., implementation of a TAN system becomes the only alternative [17, 18].

TAN approaches can be split into two categories based on the type of sensor supplying the terrain data: *passive imaging* or *active range sensing*. Passive imaging has the advantage that the sensors (cameras) are mature and easy to accommodate on a UAV. Some of the passive imaging approaches can provide navigation measurements from any altitude. However, passive imagers have the distinct disadvantage that they are ineffective in poor illumination and weather conditions which casts challenging constraints on mission planning. The active range sensors (such as radar altimeters) have the advantage that they operate under any illumination condition. However, active range sensors are less mature than passive imagers and they have limited maximum operating ranges which limit the altitude at which TAN measurements can be made available [8].

The TAN approach can also be classified based on the structure of the algorithm used to compare surface measurements to the map, *correlation approaches* and *pattern matching approaches* [18, 19]. In correlation approaches, a contiguous patch of the surface is acquired using the onboard sensor. If a passive imager is used, then the patch is essentially a subset of the image; for a range sensor, the patch is elevation map or contour. This patch is then correlated, in the image processing sense, with the onboard map. Correlation algorithms place the patch at every location in the map and then measure the similarity between the patch and the map values (this process can be visualized as raster scanning the patch across the map). If the values are similar, then the location is given a high score; if they are not, then the location is given a low score. The location in the map with the highest score is chosen as the best match location for the patch in the map. Interpolation of the correlation scores is used to obtain a sub-pixel estimate of the match position. The orientation (and altitude for imagers) is then used to compute the position of the UAV in the map coordinate system [8]. Some famous correlation TAN methods are: Image to Map Correlation for Position Estimation (e.g. DSMAC) [20] and Altimeter to DEM Correlation (e.g. TERCOM) [21].

Besides these, the main drawback of the TAN systems is that they cannot work on a flat area or over the sea. In order to find a position match, the system needs an undulating terrain or some specific terrain features like well-known craters, hills, rivers etc. When the height values which are determined from the altimeter are constant, the slope values will remain constant too. In order to attain a terrain that is rough enough to navigate on, there is a need for an automated terrain analysis system like the one described in [22] where the flight path can be automatically determined by examining the results of the smoothness-based segmentation which shows the areas in the image that surpass a degree of smoothness.

Pattern matching approaches use landmark matching instead of patch correlation. Landmarks are specific terrain locations that can be extracted from the map and also have distinct characteristics that make them amenable to comparison to other landmarks. For example, hills, rivers, craters are often used as landmarks because they can be easily extracted reliably from image data over a broad range of image scales and illumination, e.g. the diameter of the crater is an identifier that can be used for matching. The relative distances and angles between landmarks are also used during the matching procedure. Some famous pattern matching TAN methods are: Scale Invariant Feature Transform (SIFT) [23], Shape Signature Pattern Matching, Onboard Image Reconstruction for Optical Navigation (OBIRON) [24].

A range of different techniques have been developed to obtain position fixes from the comparisons of measured and database terrain heights. The TERCOM system has been successfully applied in cruise missile systems, which combines onboard radar-altimeter readings with a preloaded Digitized Terrain Elevation (DTE) map to estimate the INS errors as well as guiding the low-flying missile at a fixed height above the ground. The TERrainPROfile Matching TERPROM system correlates passive sensor data with a terrain database. It can provide terrain proximity and avoidance information as well as INS aiding capability and it has been widely adapted as a navigation system within various aircrafts [25]. TERCOM and TERPROM are such examples in which, the difference between the radar altimeter generated and database indicated terrain height is input as a measurement to Extended Kalman Filter (EKF). The principal advantage of the EKF approach is relative simplicity and comparatively low processor load. However, it relies on accurate knowledge of the terrain gradient below the aircraft, which is an over-demanding requirement. Moreover, utilization of TERCOM or TERPROM on UAVs is still an open and challenging issue.

In this study, we have developed a preliminary TAN system for UAVs, which neither necessitate an onboard passive imagery sensor and/or GPS receiver nor an accurate knowledge of the AOI terrain gradient data. Our studies show that TAN systems can easily be deployed on UAVs to be an effective auxiliary navigation system. In the following subsection the main attributes of the exemplary method is explained in detail.

3 The Exemplary TAN Algorithm

Metaheuristics are used for combinatorial optimization problems. In such problems, whether it is a minimization or maximization problem, an optimal solution is sought in a discrete search space. As most of the design problems in engineering suffer from the vast dimensionality, it is infeasible to use exhaustive search or analytical methods. Metaheuristics are also used for problems over real-valued search-spaces. Their advantage is that the function to be optimized need not be continuous or differentiable and it can also have constraints [26].

The search-space of a TAN system is also so huge that with a deterministic algorithm, providing a UAV with exact location information in a feasible amount of time is nearly impossible [9]. In addition, the miss probability of the matching the height values (that a UAV takes along its journey) with the DEM data is high because of the resolution gaps on the DEM data. These problems necessitate the usage of metaheuristic approaches for TAN systems. In this study, a metaheuristic TAN algorithm which is based on SA algorithm has been developed. The SA algorithm and the proposed technique are described briefly in following subsections:

3.1 Simulated Annealing Algorithm

SA algorithm is a commonly used metaheuristic. The main advantage of SA algorithm among other metaheuristic approaches is that it works relatively fast with avoiding to local optima [26]. It is named as 'simulated annealing' because conceptually this method is similar to a physical process known as annealing. In an annealing process, a material is heated into a liquid state then cooled back into a recrystallized solid state. Similarly, an SA algorithm starts with an initial complete feasible solution and iteratively generates additional solutions as it is annealed. Also it can exactly or approximately evaluate candidate solutions and maintain a record of the best solution obtained so far. While SA is essentially *memoryless* and a probabilistic device is used to escape from local optimum, it can be regarded as a fast and thus powerful technique.

The algorithm works as follows: at any iteration k, we have a current solution x_c and a candidate solution x from the neighborhood $N(x_c)$. As the localization of the UAV is a maximization problem, if $f(x) > f(x_c)$, then x becomes the new x_c on the next iteration. If $f(x) < f(x_c)$, then there is still a chance that x replaces x_c. The associated annealing probability can be described as follows:

$$P(x \rightarrow x_c) = exp\left[\frac{f(x) - f(x_c)}{T_k}\right] \quad (2)$$

where T_k is the annealed temperature parameter. The probability of accepting the inferior solution x is seen to decrease as the performance gap between x_c and x increases or as the temperature becomes smaller. The sequence of temperatures usually satisfies $T_1 \geq T_2 \geq \cdots$, that is, the temperature is gradually decreased. In our simulations, we have used different temperature values ranging from 100,000 to 100 and halved the temperature value at each iteration. Decreasing the T means that in the early iterations, diversification is more likely and in the later stages intensification is achieved. When the search is terminated, a local search is made in order to ensure that the final solution is at a local optimum.

3.2 Proposed Algorithm

As in our previous work of [9], the algorithm takes as input a heightmap. We assume that in advance of a flight this heightmap is preloaded into the vehicle's memory. As one of the main objectives of this study is to relieve a UAV from utilizing beacon based navigation sensors, our algorithm does not depend on any information coming from GPS, cameras etc.

On a flight, the active ranging sensor of a UAV, such as a radar altimeter, collects elevation data between two sampling intervals as shown in Fig. 2. A sampling interval is chosen to be 3–5 s for rough terrains and 5–15 s for undulating terrains. For example, if a UAV has a velocity of 40 m/s, and the frequency of the radar altimeter is 10 Hz, then within 5 s our UAV will displace 200 m and collect 50 height values which form an array of height values as illustrated in Fig. 3. We named each array of height values as a *profile*. When sufficient number of height values is collected, these values are fitted into a line with the help of minimum square roots method. The result of minimum square roots is the well-known line formula as

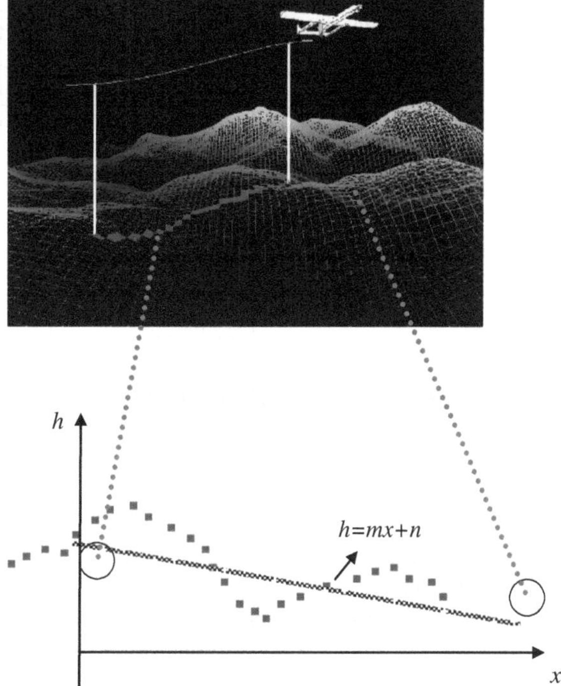

Fig. 2 Collected height values form a line which has a slope value of α

shown in (3). The slope of the profile is calculated by (4) which bear critical information about a terrain profile. Such as, a negative α value means a descending hillside and a positive value corresponds to an ascending terrain. The bigger the absolute value of α, the steeper the terrain gets.

$$h = mx + n \tag{3}$$

$$\alpha = ArcTan(m) \tag{4}$$

Afterwards, the estimated slope value is searched within the DEM file. As illustrated in Fig. 3, the projection of altimeter swaths resembles a virtual line on the map. The pseudo code of the algorithm is shown in Algorithm 1. As stated earlier, it is assumed that the UAV makes its journey on a predefined terrain so the proposed algorithm starts with loading the $n \times n$ sized DEM file into memory. The sampling number, *Sn* which corresponds to the number of the sampling points that the UAV is on and the initial heat, T, values are initialized.

Fig. 3 The UAV's flight path is virtual line on the map

25	23	36	7	0	8	9	17	19	0	36
14	45	48	36	25	33	38	20	8	25	48
5	33	5	25	14	7	14	14	25	14	5
7	14	7	48	5	25	5	5	28	5	7
36	38	36	8	7	28	7	7	35	7	36
25	49	25	49	36	35	36	36	49	36	25
38	20	8	14	17	49	36	19	14	17	8
5	34	13	23	19	5	25	12	5	19	13
7	17	14	45	48	36	25	5	7	48	14

Algorithm 1 SA based TAN system on lunar surfaces

1. **Load** *DEM(nxn);*
2. **Initialize** *M*, *Best*, *T*, *eps*;
3. *Sn* = 0;
4. **Repeat**
5. *Sn* ++;
6. *Get(x);* //get displacement from UAV;
7. *Get* (x_{slope})*;* //get the actual lidar slope from UAV
8. **For** jj = *size(M-Best)* **to** *k* **do**

   Select a random p_1, $p_1 \subset E$;//start pixel from entrance area
   Select a random β, $-45 \leq \beta \leq 45$;
   ($slope$, p_2) = Bresenham(p_1,α,x);
   update *M*;

9. **end for;**
10. **For** kk = 1 **to** *size(M)* **do**

    if ($|slope-x_{slope}| \leq eps$) & ($rand[0,1] < 1/\exp(|slope-x_{slope}|/T)$)

    update *Best*;
    update *T*;

    end if;

11. **end for;**
12. update *M with Best*;
13. **Until** satisfaction OR end of sampling;
14. **Display** Best;

The *T* values are set to be different values ranging from 10^2, 10^3, 10^4 to 10^5 in order to determine the best initial heat. At the very start of the algorithm the displacement, *x*, and the evaluated slope, x_{slope}, on the corresponding *Sn* is taken from the UAV with the *Get*() function. *x* indicates the amount of pixels that the UAV flied over and x_{slope} indicates the calculated slope in that *Sn*. While the aim of the algorithm is to find a best location match for the UAV the x_{slope} value is searched stochastically within the DEM. This search is achieved by taking a random initial pixel, *p1*, and a random flight angle, β, from the entrance area, *E*. The β values are restricted with *−45* and *45* degrees in order to maintain a direct flight from west to east and enabling unrealistic flight paths. Afterwards, the swath and the slope of the selected positions are calculated with utilizing Bresenham line drawing algorithm [27]. Initially, this procedure is iterated *k* times and all the evaluated values *(*p_1*, p2, x,* α*)* are written into memory, *M*. *M* comprises the solution space of the evaluation. By controlling all the solutions with the actual slope x_{slope} and the evaluated slope which reside within an amount epsilon of errors, *Eps*, and the annealing heat, *T*, the best results are derived with the following control:

$$\left(\left|slope - x_{slope}\right| \le eps\right) \wedge \left(rand[0,1] \le e^{-\frac{slope - x_{slope}}{T}}\right) \tag{5}$$

The best results are collected into a matrix, *Best*, which is then placed into the *M* matrix. This adaptation gives rise to escape from the local optima while keeping more adequate solutions in the solution set. After looping for a specific number of iterations, in the case that the UAV is out of the map or a satisfactory location solution is estimated within two consecutive iterations, the algorithm terminates.

4 Results and Discussion

In this study, we have utilized two types of terrain structures which are named as rough terrain and undulating terrain. These terrains are derived from a 3D DEM data of north-western part of Turkey which has 30 m resolution. By utilizing these maps, a latitude/longitude information is achieved where the values of all *(x,y)* coordinates correspond to the actual height of the terrain. An example of the terrain types that we used for this study is given in Fig. 4.

As stated earlier, this study aims to search for the best location for a UAV throughout its journey. When the size of the DEM grows, the possible routes and slope calculations grow up exponentially that exhaustive search algorithms will be in adequate to search for a possible position within a reasonable time. Hence, SA algorithm fits well for a UAV TAN system. In addition, exact algorithms will fail to provide location information when the estimated elevation values do not reside in the memory. It is a well-known fact that representing every height values of a terrain on a heightmap is impossible. Hence, interpolation and extrapolations techniques are used to fill the height value gaps between adjacent two pixels of a terrain. In real world situations, it will be a misleading assumption to expect a height value to exactly reside in the heightmap.

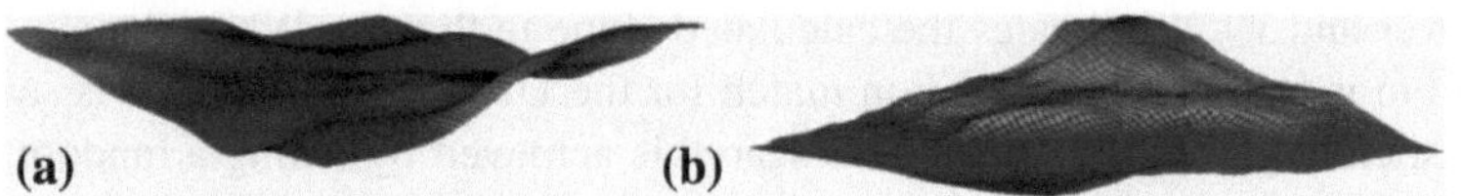

Fig. 4 Examples of terrain types used in this study **a** Undulating terrain, **b** Rough terrain

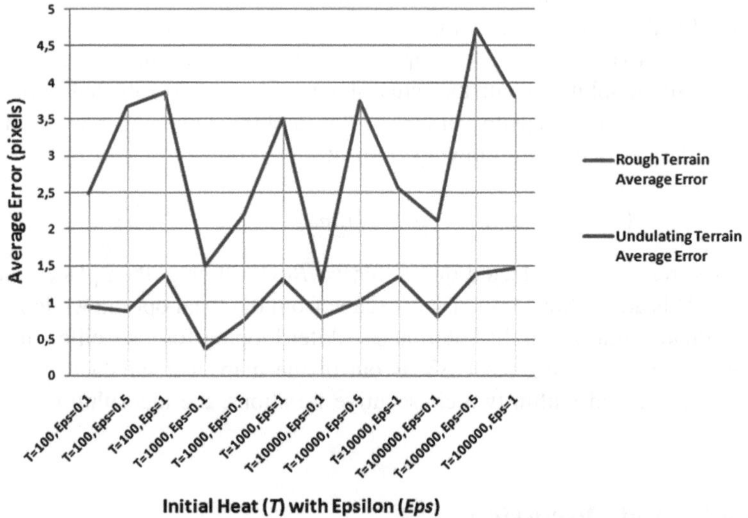

Fig. 5 Simulation results

4.1 Results of the Exemplary TAN Approach

The method followed in this study bases on SA algorithm which starts with an initial complete feasible solution and iteratively generates additional solutions. The escape from local optima is achieved by the heating parameter *T*. The *T* value is halved in each iteration and the performance evaluation of the proposed method is conducted by various initial heat values ranging from 10^2, 10^3, 10^4 to 10^5. Also the average measurement error in pixels which is expressed as *Eps* indicates the robustness of the altimeter. If *Eps* is high, the height estimation errors are also high. Throughout simulations, *Eps* is set to 0.1, 0.5 and 1.0 pixels respectively. Because the resolution of the map is 30 m, when $Eps = 0.1$ there is a ± 3 m fault range. When $Eps = 0.5$ the fault range is ± 15 m etc.

The results of the numerous simulation runs are illustrated in Fig. 5, where the vertical axes corresponds to the average location estimation error of the UAV in pixels. For example, with an initial heat of $T = 100$ and a measurement error of $Eps = 0.1$, the proposed algorithm will find a location with drifts of 0.9 and 2.5 pixels from the actual UAV position on a rough terrain and on an undulating terrain consecutively. While the map resolution is set to 30 m, 0.9 pixel drift will cause a 27 m drift from the actual position. On the other hand, a 2.5 m drift will cause a 75 m drift on an undulating terrain.

It can be inferred from Fig. 5 that the average position estimation error rates on undulating terrains are higher than the rough terrain results. This is because; the TAN methods exploit the terrain elevation structure to differentiate position estimation solutions. This phenomena result in that, TAN algorithms will fail on flat terrains and will be more successful on rough terrains.

It can also be seen from the figure that the errors decrease with decreasing *Eps* values. This is because; the correctness of the altimeter plays a crucial role on height estimations. Hence, well calibrated radar altimeters have to be chosen for TAN systems. On both terrain types, when T is set to 100 and *Eps* is set to 0.1, the minimum error rates are achieved.

4.2 Discussion and Design Objectives

In this study we have explored and experimented many of the challenges of TAN systems and proposed some design objectives which are summarized as follows:

(i) TAN systems can easily be used as an auxiliary navigation system both for military and civilian UAV missions,
(ii) In order to implement a TAN system on a UAV, barometric and radar altimeter devices have to be deployed on the UAV. The measurement sensitivity of these devices is vitally important that when the fault grows, the miss probability also grows proportionally,
(iii) The altimeters play the vital role on a TAN system. The maximum altitudes that the altimeter works properly has to be determined and fine-tuned in advance of the flight,
(iv) The AOI terrain which a UAV is flying over has to be installed on the flight computer in advance of the flight. The resolution of these maps plays an important role in finding the exact location of a UAV. With high resolution maps (which are bigger than 30 m), the miss probability also grows. In addition, the time that a UAV collects height swaths also grows which leads to delayed location estimation,
(v) With exact algorithms it is impossible to provide a UAV with location information in a feasible amount of time. On the other hand, metaheuristic algorithms (such as SA) give rise to more timely and effective solutions,
(vi) Although TAN systems are good alternatives for GPS based navigation, TAN systems cannot be operated over a totally flat terrain, lake, sea etc. hence techniques to route UAVsto more rough terrains have to be reinvented,

5 Conclusion

UAVs are effective and robust means for wide area surveillance posts. Nowadays they are regarded as "eyes on the air", in the sense that they provide wider and more powerful imagery techniques. However, they have challenging demands over autonomous navigation issues. Today, most of the UAVs are equipped with satellite based navigation instruments such as GPS. It is a well-known fact that GPS signal are prone to jamming and also they cannot be used in underwater or planetary environments. Hence, there is a need for a GPS-free navigation system which is based on the terrain information which a UAV flies over. However, development of a TAN system is a non-trivial task that development of a robust TAN system addresses many challenging issues.

In this chapter, we have examined the opportunities and challenges of TAN systems. Moreover, by developing an exemplary metaheuristic algorithm which is based on SA, we have issued some TAN system design objectives to provide a UAV with precise location information. We have shown that TAN systems can be a good alternative for traditional GPS based position estimation for UAV based surveillance missions by deploying various types of real world elevation maps. We hope that our study will reveal and enlighten many of the controversial issues in the literature, guide and encourage enthusiastic researches, and appreciated as an initial but important step on the subject.

References

1. United States Department of Defense: Dictionary of military and associated terms joint publication. 1–02, 12 Apr 2001, p 557
2. CRS Report for Congress: Homeland Security: unmanned aerial vehicles and border surveillance, 8 July 2010
3. Patrick, D., Rudol, P.: A UAV search and rescue scenario with human body detection and geolocalization. Lecture Notes in Computer Science, vol. 4830, pp 1–13 (2007)
4. Australian Centre for Field Robotics (ACFR): http://www.acfr.usyd.edu.au/research/aerospace.shtml. Accessed 05 Apr 2012
5. http://www.draganfly.com/. Accessed 05 Apr 2012
6. http://www.uasresearch.com/. Accessed 25 May 2011
7. Johnson, A.E, Montgomery, J.F.: Overview of terrain relative navigation approaches for precise lunar landing. In: Aerospace Conference, 2008 IEEE, 1–8 Mar 2008, pp. 1–10
8. Kim, J., Sukkarieh, S.: Autonomous airborne navigation in unknown terrain environments. IEEE Trans. Aerosp. Electron. Syst. **40**(3), 1031–1045 (2004)
9. Temel, S, Unaldi, N, Ince, F.: Novel terrain relative lunar positioning system using lunar digital elevation maps. In: Proceedings of the 4th International Conference on Recent Advances in Space Technologies, pp. 597–602 (2009)
10. Grewal, M.S., Weil, L.R., Andrews, A.: Global positioning systems, inertial navigation integration. Wiley, New York (2001)
11. US Department of Defense Report (2005) UAV Roadmap
12. Carroll J.: Vulnerability assessment of the transportation infrastructure relying on the global positioning system. Technical report, Volpe National Transportation Systems Center (2001)

13. Kopp, C.: Cruise missiles. Australian aviation. http://www.ausairpower.net/notices.html
14. Nygren, I., Magnus, J.: Terrain navigation for underwater vehicles using the correlator method. IEEE J Oceanic Eng **29**(3), 906–915 (2004)
15. Lewantowicz, A.H.: Architectures and GPS/INS integration: impact on mission accomplishment. In: IEEE Position, Location and Navigation Symposium, pp. 284–289 (1992)
16. Sukkarieh, S., Nebot, E.M., Durrant-Whyte, H.: A high integrity IMU/GPS navigation loop for autonomous land vehicle applications. IEEE Trans. Autom. Control **15**, 572–578 (1999)
17. Baker, W.R, Clem, R.W.: Terrain contour matching (TERCOM) premier. ASP-TR-77-61, Aeronautical systems division, Wright-Patterson Air Force Base, Aug. 1977
18. Pritchett, J.E, Pue, A.J.: Robust guidance and navigation for airborne vehicle using GPS/ terrain aiding. In Proceedings of IEEE Position Location and Navigation Symposium, pp. 457–463 (2000)
19. Adams, D., Criss, T.B., Shankar, U.J.: Passive optical terrain relative navigation using APLNav. In: IEEE Aerospace Conference, 1–8 March 2008, pp. 1–9 (2008)
20. Carr, J.C, Sobek, J.L.: Digital scene matching area correlator (DSMAC), image processing for missile guidance. In: Proceedings of the Society of Photo-Optical Instrumentation Engineers, vol. 238, pp. 36–41 (1980)
21. Golden, J.: Terrain contour matching (TERCOM): a cruise missile guidance aid. In: Image Processing for Missile Guidance. SPIE, vol. 238 (1980)
22. Rahman, Z, Jobson, J.D., Woodell, G.A, Hines, G.D. (2006) Automated, onboard terrain analysis for precision landings. In: SPIE—the International Society For Optical Engineering, vol. 6246, p 62460J
23. Johnson, A., SanMartin, M.: Motion estimation from laser ranging for autonomous comet landing. In: Proceeding International Conference Robotics and Automation (ICRA '00), pp. 132–138 (2000)
24. Gaskell, R.: Automated landmark identification for spacecraft navigation. In: Proceedings of the AAS/AIAA astrodynamics specialists conference, AAS Paper # 01–422 (2001)
25. Robins, A.: Recent developments in the 'TERPROM' integrated navigation system. In: Proceeding of the ION 44th Annual Meeting, June 1998
26. Silver, E.A.: An overview of heuristic solution methods. J. Operational. Res. Soc. **55**, 936–956 (2004)
27. Hearn, D., Baker, M.P. (1994) Computer graphics. Prentice Hall, Upper Saddle River

Automatic Target Recognition in Multispectral and Hyperspectral Imagery Via Joint Transform Correlation

Mohammad S. Alam and Adel Sakla

Abstract In this chapter, we review the recent trends and advancement on automatic target recognition (ATR) in multispectral and hyperspectral imagery via joint transform correlation. In particular, we discuss the one-dimensional spectral fringe-adjusted joint transform (SFJTC) correlation based technique for detecting very small targets involving only a few pixels in multispectral and hyperspectral imagery (HSI). In this technique, spectral signatures from the unknown HSI are correlated with the reference signature using the SFJTC technique. This technique can detect both single and/or multiple desired targets in constant time while accommodating the in-plane and out-of-plane distortions. Furthermore, a new metric, called the peak-to-clutter mean (PCM), is introduced that provides sharp and high correlation peaks corresponding to targets and makes the proposed technique intensity invariant. This technique is also applied to the discrete wavelet transform (DWT) coefficients of the multispectral and HSI data in order to improve the detection performance, especially in the presence of noise or spectral variability. Detection results in the form of receiver-operating-characteristic (ROC) curves and the area under the ROC curves (AUROC) are used to show the performance of the proposed algorithms against other algorithms proposed in the literature. Test results using real life hyperspectral image data cubes are presented to verify the effectiveness of these proposed techniques.

M. S. Alam (✉) · A. Sakla
Department of Electrical and Computer Engineering,
University of South Alabama, Mobile, AL 36688-0002, USA
e-mail: malam@usouthal.edu

A. Sakla
e-mail: asakla@usouthal.edu

Augment Vis Real (2014) 6: 179–206

DOI: 10.1007/8612_2012_5

Published Online: 11 September 2012

Keywords Hyperspectral image processing · Automatic target detection · Spectral variability · Spectral fringe-adjusted joint transform correlation · Spectral signature · Spectral variability · Wavelet transform · DWT coefficients

1 Introduction

Hyperspectral imaging (HSI) spectometry is a new technology for remote sensing and target detection applications from airborne and spaceborne platforms [1–4]. Pattern recognition is one of the fundamental tasks in HSI exploitation. Pattern recognition deals with the detection and identification of a desired pattern in an unknown input scene and also with the determination of the spatial location of the desired object if present. The performance of a pattern recognition algorithm depends on the available spatial information corresponding to the object of interest. One way to improve the detection accuracy is to introduce additional information about the object, such as the reflectance of the material at different wavelength bands i.e., spectral information. Hyperspectral sensors have been developed to provide sensor radiance spectrum corresponding to material characteristics [1]. The fundamental idea for HSI stems from the fact that, for any given material, the amount of radiation that is reflected, absorbed, or emitted depends on the wavelength. In general, hyperspectral sensors measure radiation in the 0.4–2.5 nm region of the electromagnetic spectrum, and HSI may be defined as imagery taken over many spectrally contiguous and spatially co-registered bands. HSI sensors generate images containing both spatial and spectral information that can be used in remote sensing detection and classification applications [2]. HSI sensors provide plenty of spectral information to uniquely identify materials by their reflectance spectra. Although it is theoretically possible for two completely different materials to exhibit the same spectral signature, targets in ATR applications are typically man-made objects with spectra that differ considerably from the spectra of natural background materials [3]. In contrast to multispectral sensors, which measure reflectance values at wide wavelength bands, hyperspectral sensors measure reflectance values at narrow, contiguous wavelength bands. Consequently, the richer information in HSI has better potential in ATR applications than multispectral imagery.

Automatic target recognition (ATR) is a vital and complex step in image processing and exploitation. ATR has experienced significant strides with the advent of HSI sensors. ATR systems should be able to detect, classify, recognize, and/or identify targets in an environment where the background is cluttered and targets are at long distances and may be partially occluded, degraded by weather, or camouflaged [5]. The goal of spectral target detection is to identify pixels containing a material whose spectral composition is known. In HSI target detection applications, the targets are present sparsely throughout an image, which may account for less than 1 % of the total pixels in an input scene, rendering traditional spatial processing techniques impractical. Consequently, most HSI detection algorithms exploit the spectral information of the scene, an approach otherwise known as *nonliteral exploitation* in the HSI literature [2]. One of the main challenges in HSI processing is *spectral variability*, which refers to the phenomenon that spectra observed from samples of the same material will never be identical. While several detection algorithms have been developed over the years, spectral variability remains a major challenge for these algorithms [2, 3, 6].

In HSI, the target is usually present in the form of a few pixels which makes it difficult to get the maximum likelihood estimate of the desired class. Various statistical technique based detection algorithms have been proposed in the literature for HSI applications using hypothesis testing [3]. Deterministic approaches do not exploit the statistical information of the targets and background classes and the decision is based on some criteria such as the angle between two spectral vectors or the correlation peak intensity. In statistics, it is well known that when the two distributions are multivariate normal, then by using the likelihood ratio test, unequal covariance matrices lead to quadratic discriminant analysis, such as the Neyman-Pearson detector [4]. For this type of detectors, it is necessary to have a distribution model for both target and background classes. In general, targets constitute a few pixels in HSI. Accordingly it is difficult to estimate the target class distribution parameters. Therefore, in some cases, it is assumed that the covariance matrices of both classes are the same allowing linear discriminant analysis or Fisher discriminant analysis. However, this assumption may not produce reliable results if the covariance of the target class is completely different from the covariance of the background [7, 8].

Anomaly detectors are widely used for hyperspectral detection applications due to the inherent fluctuations in spectral signatures [3]. Anomaly detection was proposed by Reed and Xiaoli (RX) [9], where detection decision is based on the maximum likelihood estimates of the background class only. The RX algorithm involves a practical detector, which does not rely on a priori information about the target. The RX algorithm calculates the Mahalanobis distance of a pixel vector from the mean of the background class, and then classifies it either as background or as a target using a suitable threshold. The main challenges of RX algorithm are modeling the background statistics, and setting the precise threshold level for a given constant false alarm rate.

Among the deterministic algorithms, spectral angle mapper (SAM) is commonly used for hyperspectral image processing. If the reference spectral signature is available in the form of reflectance spectra, SAM algorithm can be applied for detection purposes. The SAM algorithm determines the similarity of an unknown spectral signature to a known reference spectrum [10]. In this algorithm, the cosine of the angle between the test and reference spectra is evaluated and it provides good results only for targets having well separated distributions with small dispersions [3]. One of the advantages of the SAM algorithm is that it does not require any training step. However, this algorithm may not provide robust results in the presence of noise and variations of the spectral signatures.

The JTC technique has been found to yield excellent correlation output for two-dimensional and three-dimensional pattern recognition applications [11–14]. Among the various JTC techniques, the fringe-adjusted (FJTC) technique has been found to yield significantly better correlation output compared to alternate JTC algorithms [15–19]. In this section, a new spectral fringe-adjusted JTC (SFJTC) based detection algorithm for HSI is proposed and evaluated to address the aforementioned problems. Input spectral signatures from a given hyperspectral image data cube are correlated with the reference signature using the SFJTC technique. The main advantage of this technique over statistical techniques is that it does not require any training at any step of the detection process. In statistical techniques, although the reference signature may be known, it may not be possible to estimate the covariance of the target class. In the proposed technique, the detection performance is sensitive to the shape but not to the amplitude of the reference signature, which makes the proposed algorithm superior because the signature (shape) of a material is usually preserved whereas the intensity may change due to environmental conditions. This technique provides sharp and high correlation peaks for a match and negligible or no correlation peaks for a mismatch. In addition, the SFJTC technique is also applied to the discrete wavelet transform (DWT) coefficients of the HSI data in order to improve the detection performance, especially in the presence of noise or spectral variability.

A variation of the SFJTC algorithm is known as the spectral joint fractional Fourier transform correlation (SJFRTC), where the fractional Fourier transform correlation (FRTC) is applied instead of the full Fourier transform correlation [20]. The post-processing step uses a binary differential operation to enhance its performance, resulting in delta-function-like correlation peaks without dc interference. The PCM measure is applied instead of absolute correlation peak value to make the performance independent from the intensity variation of the spectral signature.

2 Spectral Fringe-Adjusted Joint Transform Correlation

The joint transform correlation (JTC) technique has been found to be an effective tool for two- and three-dimensional pattern recognition applications [11]. Among the various JTC techniques, the FJTC technique has been found to yield better correlation output [15–18]. To achieve efficient detection of challenging objects in HSI, such as single pixel objects, we propose herein the SFJTC technique.

When the input scene contains only one spectral signature, the correlation output produced in a JTC technique includes three terms, a strong zero-order term at the center flanked by a pair of cross-correlation terms in the correlation plane [11, 12, 21]. The FJTC technique provides enhanced correlation performance by incorporating a real-valued filter, called the fringe-adjusted filter (FAF), leading to delta-function-like correlation peaks for a match [15].

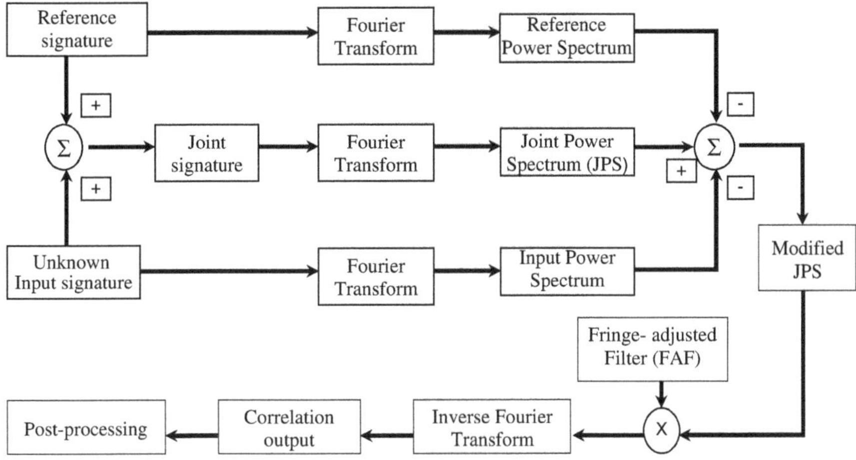

Fig. 1 SFJTC algorithm block diagram

2.1 SFJTC Analytical Modeling

The block diagram of the proposed SFJTC technique is shown in Fig. 1, where the reference spectral signature $r(x)$ and the input spectral signature $t(x)$ are separated by a distance of $2x_0$ along the x-axis [22]. The input joint spectral signature $f(x)$ can be expressed as

$$f(x) = r(x + x_0) + t(x - x_0) \tag{1}$$

Applying Fourier transform to Eq. (1), we get

$$F(u) = |R(u)| \exp [j\phi_r(u)] \exp (jux_0) + |T(u)| \exp[j\phi_t(u)] \exp(-jux_0) \tag{2}$$

where $|R(u)|$ and $|T(u)|$ are the amplitudes, and $\phi_r(u)$ and $\phi_t(u)$ are the phases of the Fourier transforms of $r(x)$ and $t(x)$, respectively, and u is a frequency-domain variable. The joint power spectrum (JPS) corresponding to Eq. (2) is given by

$$\begin{aligned} |F(u)|^2 = |R(u)|^2 + |T(u)|^2 + |R(u)||T(u)|^* \exp [j\{\phi_r(u) - \phi_t(u) + 2ux_0\}] \\ + |R(u)|^* |T(u)| \exp [j\{\phi_t(u) - \phi_r(u) - 2ux_0\}] \end{aligned} \tag{3}$$

In Eq. (3), the first two terms correspond to the zero-order terms and the last two terms correspond to the desired cross-correlation between the known reference spectral signature and the unknown input spectral signature.

To eliminate the zero-order terms, the Fourier plane image subtraction technique [16–18] is used, where the input-signal-only power spectrum and the reference-signal-only power spectrum are subtracted from the JPS to yield the modified JPS [22, 23], expressed as

$$\begin{aligned}|I(u)|^2 &= |F(u)|^2 - |R(u)|^2 - |T(u)|^2\\ &= |R(u)||T(u)|^* \exp\left[j\{\phi_r(u) - \phi_t(u) + 2ux_0\}\right]\\ &\quad + |R(u)|^*|T(u)| \exp\left[j\{\phi_t(u) - \phi_r(u) - 2ux_0\}\right]\end{aligned} \quad (4)$$

The classical JTC technique yields large correlation side-lobes, large correlation peak width, and strong zero-order peak [15–17], which deteriorates the detection performance. To provide sharp correlation peaks and low correlation side lobes, the FJTC technique is used, where the modified JPS found in Eq. (4) is multiplied by the real-valued fringe-adjusted filter (FAF) before applying the inverse Fourier transform operation to yield the correlation output [15]. The FAF is characterized by the transfer function defined as [17]

$$H(u) = \frac{A(u)}{B(u) + |R(u)|^2} \quad (5)$$

where $A(u)$ and $B(u)$ are either constants or functions of u. When $A(u) = 1$ and $|R(u)|^2\rangle\rangle B(u)$, the FAF can be approximated as

$$H(u) \approx \frac{1}{|R(u)|^2} \quad (6)$$

The modified JPS of Eq. (4) is then multiplied by the FAF of Eq. (5) to generate the fringe-adjusted JPS, given by

$$G(u) = H(u) \times |I(u)|^2 \quad (7)$$

Finally, an inverse Fourier transform of the fringe-adjusted JPS yields the correlation output, given by

$$c(x) = F^{-1}\left\{H(u) \times |I(u)|^2\right\} \quad (8)$$

2.2 Post-Processing

In case of hyperspectral image datasets, one deals with n-dimensional pixel vectors, $x = [x_1, x_2, \ldots x_n]^T$, where each pixel vector undergoes the correlation process described in Sect. 2.1. If one records the highest correlation peak to identify the location of an input signature, it may not be possible to obtain distinguishable correlation performance since the highest peak values in the correlation plane with true signature and those with false signature often do not differ significantly. For demonstration purposes, at first we investigated the performance of the proposed technique using simple reference and target signatures as shown in Fig. 2. Figure 2 shows the known reference spectral signature, an unknown input spectral signature

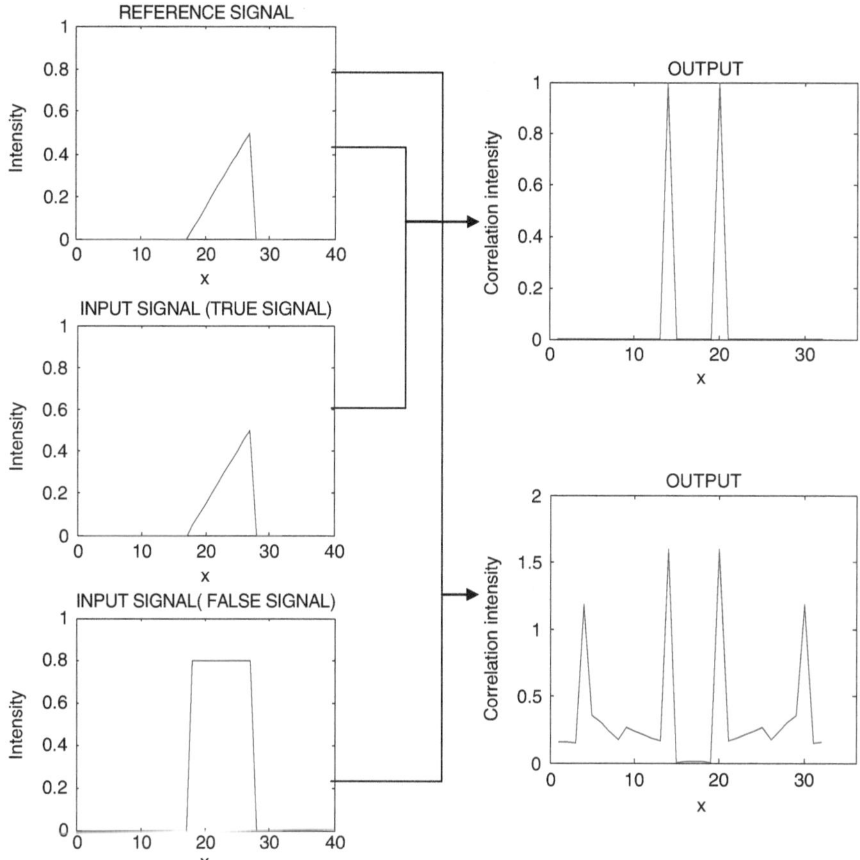

Fig. 2 Correlation results with ordinary signals

which is the same as the reference signal (true signal), and another input spectral signature (false signal) which has higher intensities then the reference signal, and the corresponding correlation outputs. From the correlation output of Fig. 2, we observe that the highest peak generated by the SFJTC with the true input signal is less then the highest peak generated with the false signal. This is due to the dependence of all JTC techniques on the intensity of a signal. The false input signature has a different shape but has higher intensity. Therefore, when the false input signal and the reference signal are correlated, it is possible to get a higher correlation peak leading to false target detection.

To alleviate this problem, in this section, a novel metric, called the peak-to-clutter mean (PCM) is introduced. The PCM metric [19] is used to identify the location of an input spectral signature, defined as

Table 1 Correlation peak intensity (CPI) and peak-to-clutter mean (PCM) for the true and false signals shown in Fig. 2 generated by the Spectral FJTC technique

Input signal	CPI	PCM
True signal	1.00	726234106.38
False signal	1.50	6.77

$$PCM = \frac{CPI}{\frac{2}{m}\sum_{i=1}^{\frac{m}{2}} c(i)} \tag{9}$$

where *CPI* represents the correlation peak intensity corresponding to the highest point in the correlation output, and *m* represents the dimension of the joint spectral signature.

The correlation results obtained using the SFJTC technique is shown in Table 1. From Table 1, it is evident that the ratio of PCM is significantly higher than the ratio of the highest peaks between the true and false spectral signatures. Moreover, the PCM remains the same even if the reference reflectance at all spectral bands drops or increases by some constant $k > 0$, thereby making this method sensitive to spectral shape rather than the intensity which enables the SFJTC to yield a delta-function-like correlation peak for a match. To illustrate this concept, we correlated a known reference spectral signal with two unknown input spectral signatures corrupted by noise and other artifacts as shown in Figs. 3a and d, respectively. Figure. 3a shows that the unknown input spectral signal is almost similar to the known reference signal. The unknown input spectral signature shown in Fig. 3b contains both Gaussian noise and distortion due to the mixing of background radiance at the sensor. This background effect has been simulated by multiplying the reference spectral signature by the proportion matrix (i.e., abundances) such that a pixel contains 80 % of the reference pixel and the remaining 20 % comes from the background for that pixel. The correlation output corresponding to the spectral FJTC and spectral JTC for the above mentioned signals are shown in Figs. 3b, c, e, and f, respectively. From Fig. 3, it is obvious that despite the noise and intensity variations, SFJTC yields high PCM and detects the desired target without any ambiguity. The numerical data for the above mentioned PCM and CPI metrics for these two techniques under various scenarios is shown in Table 2. From Table 2, we observe that the seperability factor of SFJTC with true input and false spectral signatures is almost 15. However, no essential difference is evident when the conventional JTC technique is used.

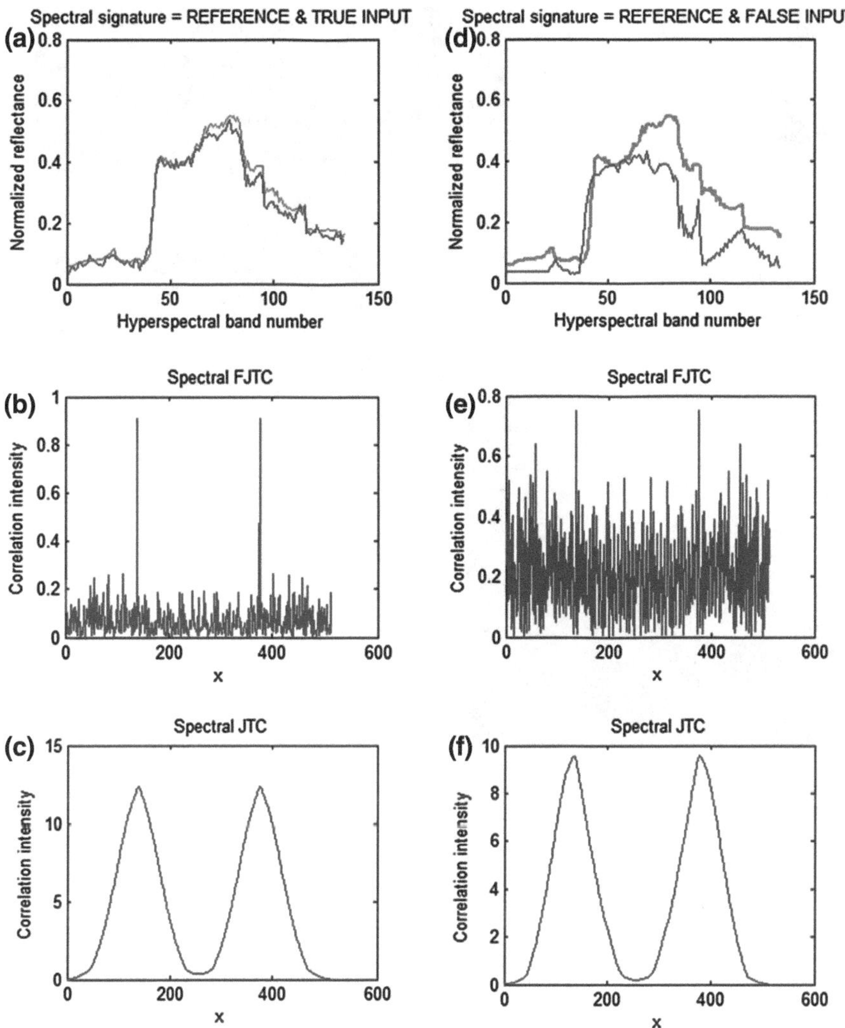

Fig. 3 Correlation results with spectral signatures

Table 2 Correlation peak intensity (CPI) and peak-to-clutter mean (PCM) for the spectral signatures shown in Fig. 3 generated by the Spectral FJTC and JTC techniques

Spectral signatures	Metric	Spectral FJTC	Spectral JTC
Reference & Reference	CPI	0.99	13.18
	PCM	9271544.42	6.97
Reference & True input	CPI	0.91	12.36
	PCM	190.67	7.04
Reference & False input	CPI	0.75	9.57
	PCM	13.05	7.65

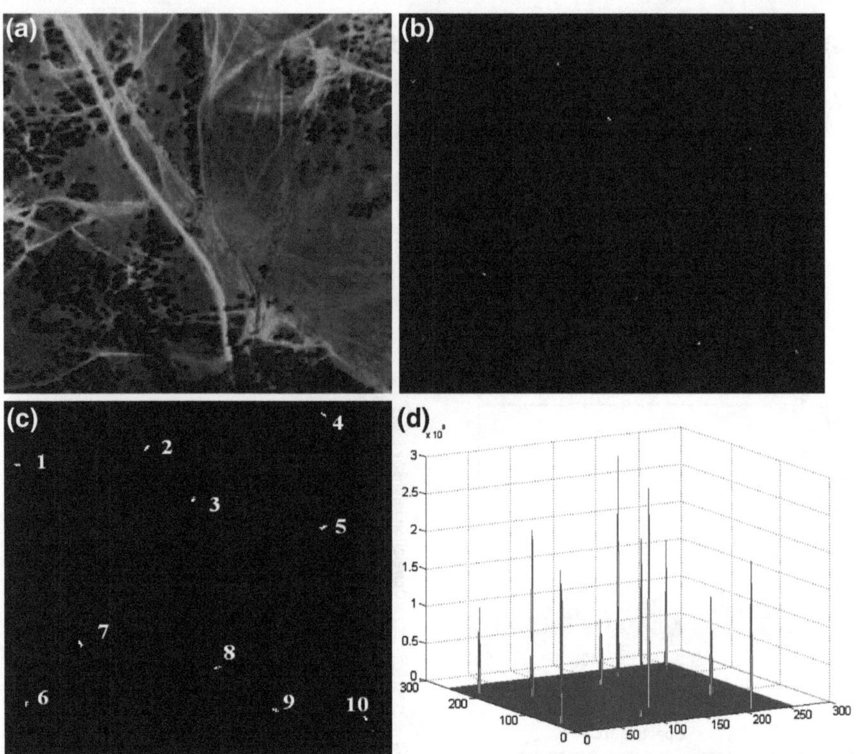

Fig. 4 Performance of SFJTC using Dataset1. **a** Dataset1. **b** All 10 targets are detected. **c** Truth mask. **d** SFJTC correlation output

2.3 Test Results

The performance of the proposed SFJTC technique has been tested using two real life hyperspectral datasets i.e., Dataset1 and Dataset2, respectively. The calibrated Dataset1 consists of 210 bands with a spectral resolution of about 10 nm. Some of these bands were in the water absorption region containing sensor errors and were removed to avoid unnecessary errors. The calibrated Dataset2 consists of 36 bands ranging from 433.7 nm to 965 nm.

Figure 4 shows the performance of the proposed SFJTC technique with Dataset1. Figure 4a shows one band from Dataset1 which contains 10 targets as shown in the truth mask of Fig. 4b. From the correlation output shown in Fig. 4c, d, it is evident that the proposed technique successfully detects all of the targets. The performance of the proposed technique was then tested with Dataset2. Figure 5a shows one band from Dataset2 which contains 11 targets as shown in the truth mask of Fig. 5b. From the correlation output shown in Fig. 5c and d, it is evident that the proposed SFJTC technique successfully detects all of the 11 targets.

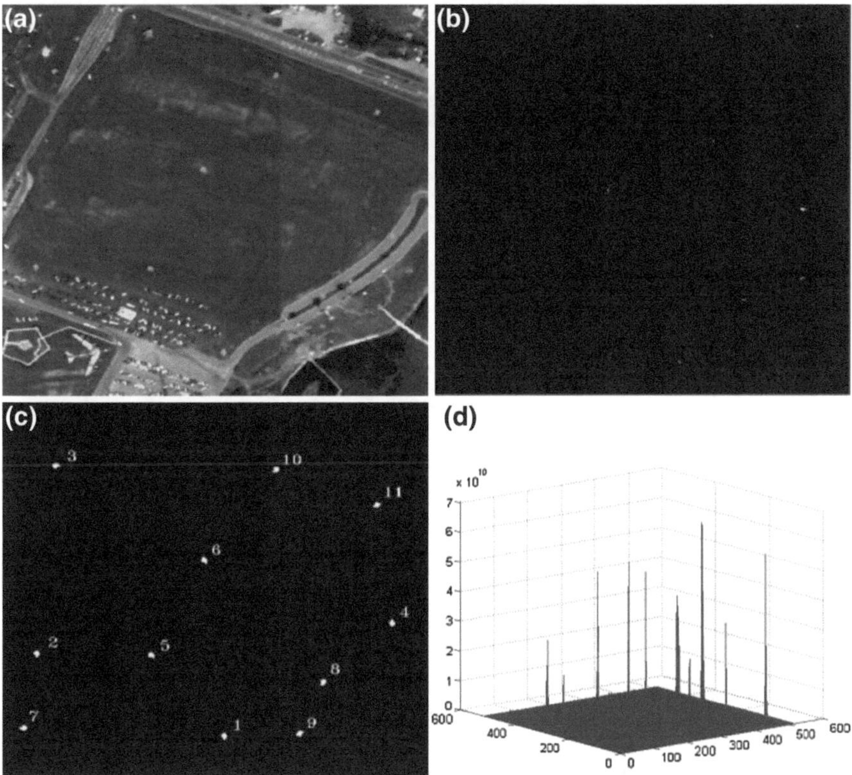

Fig. 5 Performance of SFJTC using Dataset2. **a** Dataset2. **b** All 11 targets are detected. **c** Truth mask. **d** SFJTC correlation output

For any given detector, the trade-off between the probability of detection (P_d) and false alarm (FA) rate is described by the receiver operating characteristic (ROC) curve, which corresponds to a plot of $P_d\,(\eta)$ versus $FA\ \ (\eta)$ as a function of all possible values of the threshold η [19]. Therefore, ROC curves provide the means to evaluate the performance of a detector. To compare the performance of the SFJTC algorithm with alternate techniques, such as the RX and SAM algorithms, we generated ROC curves for all of these algorithms using the same datasets. Figure 6 illustrates the ROC curves of SFJTC, RX and SAM algorithms for Dataset1. From Fig. 6, it is obvious that the SFJTC technique shows excellent correlation output and it is possible to achieve a high probability of detection while keeping the probability of false alarms at a low level. Furthermore, even for different values of threshold η, the detection rate is high confirming the robustness of the proposed technique. Figure 7 illustrates the ROC curves of SFJTC, RX and SAM algorithms for Dataset2. From Fig. 7, it is obvious that the SFJTC algorithm demonstrates similar characteristics as RX but outperforms the SAM algorithm.

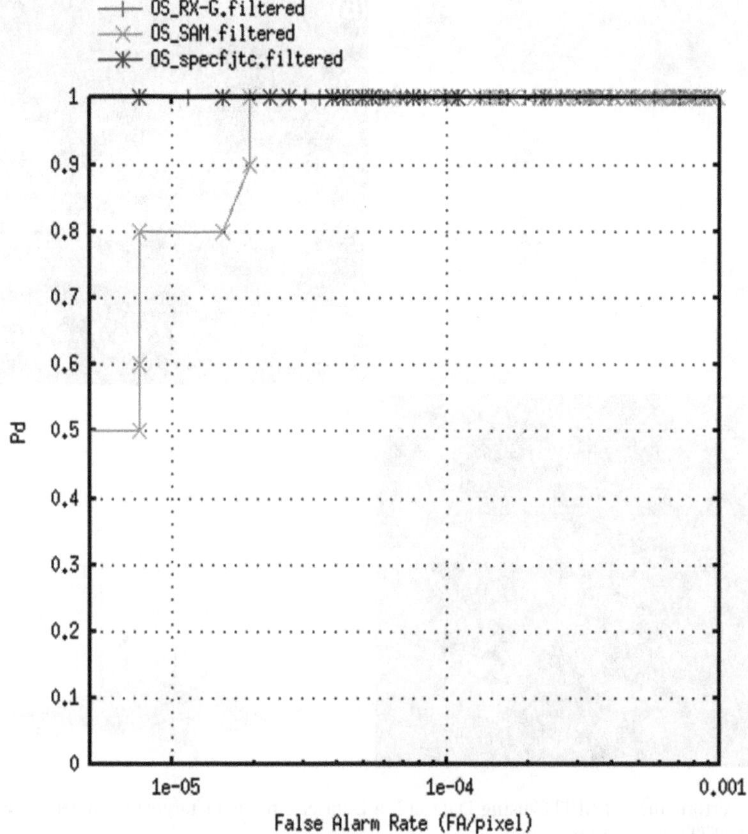

Fig. 6 Comparison of RX, SAM and Spectral FJTC algorithms with Dataset1 using ROC curve

The computational complexity of spectral FJTC is $O(2MNm\log_2 m)$ where M and N are the spatial dimensions of the image, and m represents the dimension of the canvass with two spectral signatures. The processing time in Matlab (Version 7.1) using a desktop computer with a 3 GHz processor and 4 GB RAM has been found to be 18.32 s. The proposed technique yields an independent correlation output for each pixel of the hyperspectral data cube. The processing can be enhanced by using suitable optoelectronic architectures or by parallel processing of the pixel data.

3 Conclusion

In this section, a detection SFJTC technique is proposed which exploits the spectral information of hyperspectral images to detect very small objects involving only a few pixels. Input spectral signatures from the hyperspectral image are

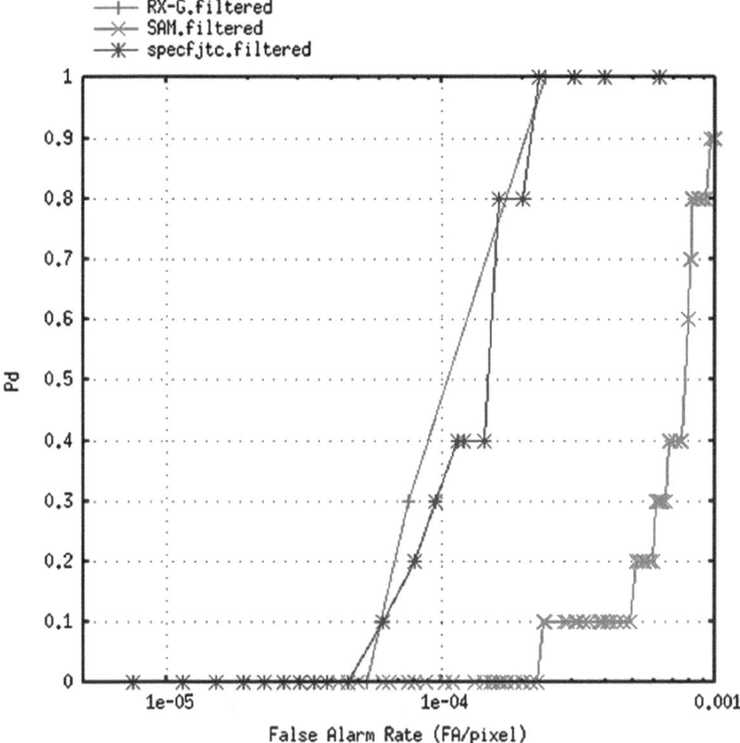

Fig. 7 Comparison of RX, SAM and Spectral FJTC algorithms with Dataset2 using ROC curve

correlated with the known reference signature using the proposed SFJTC technique. It records the ratio of the highest peak to the clutter mean for the pixel under analysis, which makes the algorithm sensitive to the shape of the reference signature but not to its intensity. The proposed technique provides excellent correlation output with sharp peaks compared to alternate detection techniques.

4 Discrete Wavelet-Based Spectral Fringe-Adjusted Joint Transform Correlation

One of the main challenges in HSI processing is *spectral variability*, which refers to the phenomenon that spectra observed from samples of the same material will never be identical. The 1-D SFJTC technique described in Sect. 2 has been used effectively for performing deterministic target detection in HS imagery. However, experiments show degraded performance when significant spectral variability is present in the target signatures [24]. In this section, we focus on the application of DWT coefficients as features for HSI target detection using the SFJTC technique. We devise a supervised training algorithm that selects an optimal set of DWT coefficients from a three-level DWT decomposition of the data to address spectral variability.

4.1 Wavelet Transform

The wavelet transform has been developed in both the continuous and discrete domains - the continuous wavelet transform (CWT) and discrete wavelet transform (DWT), respectively. Wavelets have had many successful applications in data compression [25], noise removal [26], and texture classification [27]. In the signal processing arena, they are commonly used to represent a signal in terms of its global features, yielding the general shape of the signal, and its local features, yielding the details that make the signal unique.

Wavelets have also been used in the exploitation of HSI data. Bruce and Li [28] have investigated the feasibility of derivative analysis of hyperspectral signatures for computing space-scale images and spectral fingerprints. The application of wavelet-based feature extraction for the classification of agricultural HSI has been studied by Bruce et al. [29]. They show that the wavelet transform approach is superior to conventional feature extraction methods in terms of overall classification accuracy. The use of the DWT for dimensionality reduction of HSI data has been investigated by Kaewpijit et al. [30]. They show that the DWT is superior to principal components analysis (PCA) for dimensionality reduction and yields better or comparable classification accuracy on HSI data, in addition to being more computationally efficient than PCA. Bruce et al. [31] investigate the use of the wavelet coefficients' scalar energies as features for use in a statistical classification system of hyperspectral signals, particularly focusing on the ability to classify subpixel targets.

The DWT is a highly efficient alternative to the CWT that works by discretizing the scale and translation parameters of the CWT. A computationally fast implementation of the wavelet transform is known as the *Mallat* algorithm, which represents the wavelet basis functions with a pair of low-pass and high-pass filters that meet certain constraints [32–34]. The general form of the 1-D DWT is shown in Fig. 8.

The original signal $f(n)$ is passed through low-pass and high-pass filters with FIRs $h[n]$ and $g[n]$, respectively. The filtered signals are then decimated by a factor of two, yielding the coefficients at the first level of decomposition. At each decomposition level, the outputs along the low-pass branch are known as the *approximation* coefficients, while the outputs along the high-pass branch are known as the *detail* coefficients. Multiple levels of decomposition are executed by iteratively repeating the filtering and dyadic decimation procedure on the approximation coefficients. The process is repeated K times, and the approximation coefficients cA_K and detail coefficients $cD_j, j \in \{1, \ldots, K\}$ are known as the wavelet coefficients. The coefficients are often concatenated into a single vector and denoted by w:

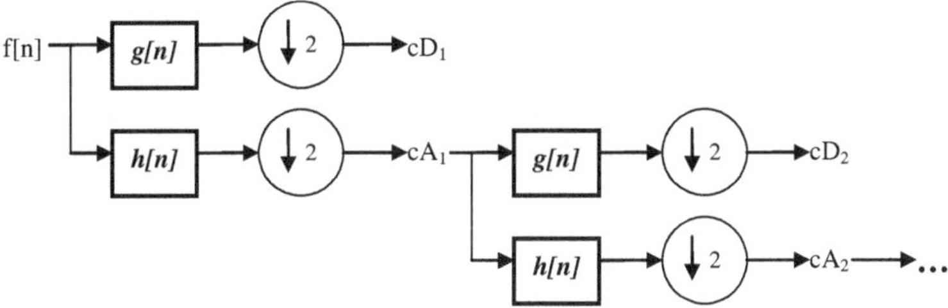

Fig. 8 Recursive filtering diagram of the Mallat algorithm for 1-D DWT

Fig. 9 CASI urban (*left*) and vegetative (*right*) scenery

$$w = [cA_K \, cD_K \, cD_{K-1} \ldots cD_1]. \tag{10}$$

Many different families of wavelets with varying properties are defined [34]. In this section, we use the well-known *Daubechies* wavelets of order 4 (*db4*).

4.2 Experimental HSI Data

Two of the data cubes in our experiments have been acquired using a CASI [35] sensor that produces 36 spectral bands ranging from 433 nm to 965 nm with a spectral resolution of 15 nm. The first data cube is urban scenery with a spatial resolution of 200 × 200, and the second data cube is vegetative scenery with a spatial resolution of 200 × 200. Visual range images of both scenes are shown in Fig. 9.

Two other data cubes have also been acquired using a HYDICE [36] sensor. This sensor operates in the Visible to Near Infrared (VNIR) and Short-Wave Infrared (SWIR) range of 400 to 2500 nm and is sampled to yield 203 spectral bands. As before, we have both urban and vegetative scenes with spatial resolutions of 200 × 200 whose visual range images are shown in Fig. 10 The low SNR

Fig. 10 HYDICE urban (*left*) and vegetative (*right*) scenery

and water absorption bands have been removed, leaving a total of 172 bands and 193 bands for the urban and vegetative scenes, respectively.

In practice, it is difficult to obtain ground-truthed HSI imagery that contains multiple real targets in low-probability target detection scenarios. The alternative is to insert simulated targets in real HSI imagery. We have randomly selected 200 pixels from each of the scenes and replaced their signatures with "corrupted" spectral signatures of a particular target material. The randomly selected locations of the targets are fixed for all four scenes. Figure 11 is the binary truth image showing the locations of the inserted targets. Note that 200 pixels account for only half a percent of the 40000 total scene pixels in each data cube, thus simulating a low-probability target detection scenario.

The targets are generated according to a first-order Markov-based model discussed in the next section. To account for the possibility of mixed pixel spectra corresponding to the boundaries of targets in the image, twenty (10 %) of the signatures have been linearly mixed in abundance ratios from 50 to 95 % using the linear mixing model [37].

4.3 Signature Generation Via Markov Model

To generate the targets that will be inserted into the scenes and simulate their spectral variability, we have used a first-order Markov-based model, defined as

$$x \sim N_K[t, \Gamma]. \tag{11}$$

In (11), x is a generated target signature, t is the pure target signature, Γ is the covariance matrix, and K is the number of bands in the HS image. It is well known that adjacent bands in a hyperspectral image are highly correlated. We utilized this

Fig. 11 Binary truth image showing locations of inserted targets

observation in conjunction with a first-order Markov model to define the covariance matrix Γ in (11) defined as

$$\Gamma = \sigma^2 \cdot R, \tag{12}$$

where **R** is a Toeplitz correlation matrix defined according to the first-order Markov model [38], given by

$$R = \begin{bmatrix} 1 & \rho & \rho^2 & \cdots & \rho^{K-1} \\ \rho & 1 & \rho & \rho^2 & \vdots \\ \rho^2 & \rho & 1 & \rho & \rho^2 \\ \vdots & \rho^2 & \rho & 1 & \rho \\ \rho^{K-1} & \cdots & \rho^2 & \rho & 1 \end{bmatrix}. \tag{13}$$

In (13), σ^2 represents the variance used to control the level of variability in the generated target class signatures. The variance σ^2 that has been used in our experiments has been varied to achieve signal-to-noise ratios (SNR) of 8, 10, 12, and 15 dB, respectively. The SNR is defined herein as the RMS of the pure target signature divided by the standard deviation of the noise, expressed as

$$SNR = \frac{\sqrt{\frac{1}{K}\sum_{i=1}^{K} t_i^2}}{\sigma}. \tag{14}$$

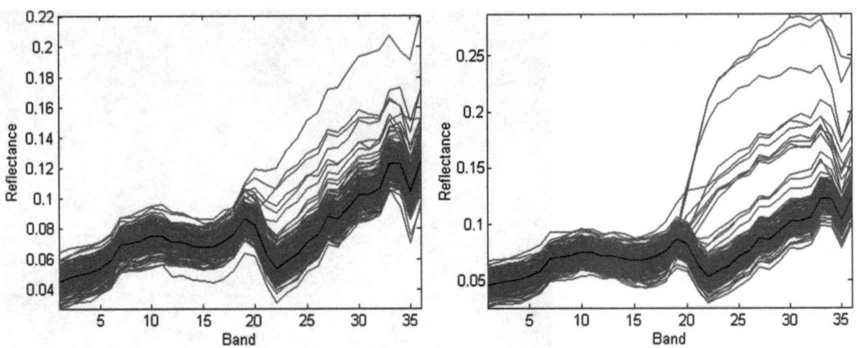

Fig. 12 Targets inserted into *CASI_urban_10* (*left*) and *CASI_veg_10* (*right*) scenes

Thus, from the original scene, four data cubes were generated with each containing corrupted targets with a specific SNR. Hence, each data cube contains a different level of target spectral variability ranging from light (15 dB), to moderate (12 dB), to heavy (10 dB), and to severe (8 dB) spectral variability. In this section, the data cubes are designated by the acquisition sensor, the type of scenery, and the SNR of the targets that have been inserted into them. For instance, for the urban scene acquired using the HYDICE sensor where we have inserted targets with a SNR of 12 dB, we referred to this data cube as *HYDICE_urban_12*.

In (13), ρ is the first-order correlation constant that we have estimated using the correlation coefficients between adjacent bands of the scene. Because the only information we have concerning the target class is the pure target signature t, we adopted an adaptive approach that uses the scene to estimate ρ, which is computed using the following steps:

1. Compute the correlation coefficient matrix of the data cube, yielding a K x K matrix.
2. Generate a vector **r** that contains the correlation coefficients between band j and the adjacent band $j + 1$. Hence, **r** has a length of K-1.
3. Estimate ρ by computing the mean of **r**.

Intuitively speaking, the reason behind using a first-order Markov model to generate the covariance matrix is that the reflectance values at b and j will closely resemble the reflectance values at b and j-1 due to the increased spectral resolution of the HSI sensor. Figures 12 and 13 show targets with a SNR of 10 dB that have been inserted into the scenes using our first-order Markov-based Gaussian model. Notice the subtle differences between some of the generated targets for each sensor due to the signatures that have been mixed with the background signatures of the respective scenes.

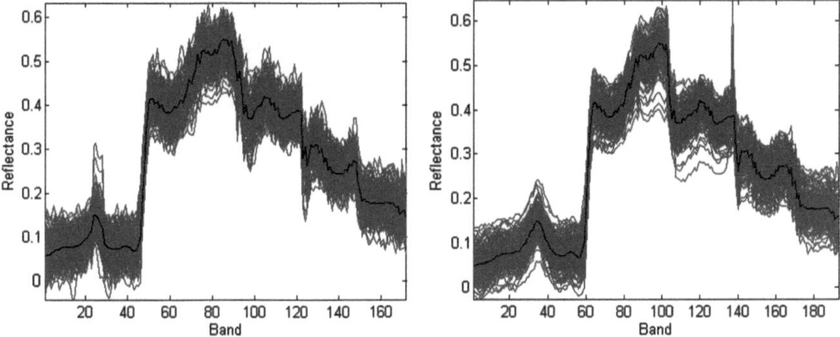

Fig. 13 Targets inserted into *HYDICE_urban_10* (*left*) and *HYDICE_veg_10* (*right*) scenes

Table 3 DWT coefficient combinations for decomposition levels 1–3

Level	DWT coefficient combinations
1	cA_1, cA_1cD_1
2	cA_2, cA_2cD_1, cA_2cD_2, $cA_2cD_2cD_1$
3	cA_3, cA_3cD_1, $cA_3cD_1cD_2$, $cA_3cD_1cD_3$ cA_3cD_2, cA_3cD_3, $cA_3cD_3cD_2$, $cA_3cD_3cD_2cD_1$

4.4 DWT-Based SFJTC

In this section, we present a supervised training algorithm that will determine the best set of DWT coefficients to use for SFJTC-based detection on a scene. Table 3 shows the various combinations of DWT coefficients for decomposition levels 1–3. It should be noted that a large number of DWT combinations exist with respect to the combinations of *individual* coefficients. The combinations in Table 3 are simply the various possibilities of grouping the contiguous subsets of *approximation* and *detail* coefficients at a particular decomposition level. For a given scene and the associated target signature, the DWT coefficients are generated using the *db4* wavelet at the desired decomposition level *j*.

In an earlier work [24], we investigated the efficacy of using the DWT coefficients for SFJTC-based detection with target signatures that were generated according to a Gaussian model. In the Gaussian model, target signatures are generated according to the model in (11). However, the covariance matrix is diagonal with $\Gamma = \sigma^2 I$, thus not accounting for any spectral band-to-band correlation. As shown in Ref. [24], each combination from Table 3 was used to perform detection on HS imagery and the AUROCS from detection were recorded. The results indicated that particular combinations of DWT coefficients performed better than others. Results on this data illustrate that performing SFJTC-based

Table 4 Optimal DWT coefficient results on CASI scenery

Data cube	Optimal DWT combination	AUROC using DWT combination	AUROC using original signatures
CASI_urban_15	$cA_3cD_3cD_2cD_1$	0.9987	*0.9980*
CASI_urban_12	$cA_3cD_3cD_2cD_1$	0.9987	*0.9980*
CASI_urban_10	$cA_3cD_3cD_2cD_1$	0.9987	*0.9980*
CASI_urban_8	$cA_3cD_3cD_2cD_1$	0.9974	*0.9967*
CASI_veg_15	$cA_3cD_3cD_2cD_1$	0.9917	*0.9916*
CASI_veg_12	cA_3cD_2	0.9882	*0.9916*
CASI_veg_10	$cA_3cD_3cD_2cD_1$	0.9917	*0.9916*
CASI_veg_8	$cA_3cD_3cD_2cD_1$	*0.9917*	*0.9916*

detection using the DWT coefficients leads to improved detection performance, with amounts of improvement that were proportional to the severity of spectral variability present in the input scene. In fact, AUROC values exhibited as much as 58 % improvement for scenes containing targets with significant spectral variability [24].

Here, we generate a training set consisting of target and background signatures. This target set will be used in our supervised training procedure for selecting the best set of DWT coefficients from Table 3. Concerning the target class, we are only given a pure target signature from a library; we do not have samples that characterize its spectral variability. We introduced spectral variability into the target class by generating *100* signatures according to our first-order Markov model presented in Sect. 5. We set the value of σ^2 in (12) to achieve a SNR of 10 dB to ensure sufficient spectral variability. For the background class, we can safely use random samples from the scene since targets occur with such low probability. Assuming no a priori knowledge of the scene, we randomly selected 8000 pixels, 20 % of the total pixels, for use as our background training samples. These 8100 signatures form our training set for the supervised coefficient selection process. The ratio of 100 target signatures to 8000 background signatures was chosen to mimic a low probability scenario in the training data. The SFJTC algorithm was run between the training set signatures and the pure target signature using each of the DWT combinations in Table 3. The DWT coefficient combination yielding the largest AUROC is selected as the optimal combination for a given scene.

The selected optimal DWT coefficient combinations for all data cubes are shown in Tables 4 and 5, respectively. For each data cube, the optimal DWT coefficient combination is listed along with the AUROC detection results using both the original signatures and the optimal DWT coefficient combination. From Tables 2 and 3, SFJTC-based detection using the original signatures yields mean AUROC values above 0.995. Regardless, use of the SFJTC technique with the optimal DWT combination yields slightly increased or identical AUROC values on all the scenery.

Table 5 Optimal DWT coefficient results on HYDICE scenery

Data cube	Optimal DWT combination	AUROC using DWT combination	AUROC using Original signatures
HYDICE_urban_15	$cA_3cD_1cD_3$	1.0000	1.0000
HYDICE_urban_12	$cA_3cD_1cD_3$	1.0000	0.9999
HYDICE_urban_10	$cA_3cD_1cD_3$	0.9999	0.9992
HYDICE_urban_8	cA_2cD_1	0.9752	0.9715
HYDICE_veg_15	cA_1	1.0000	0.9986
HYDICE_veg_12	cA_1	1.0000	0.9986
HYDICE_veg_10	cA_1	1.0000	0.9986
HYDICE_veg_8	cA_1cD_1	0.9995	0.9986

4.5 Results and Comparisons

The SFJTC detection algorithm using both the original signatures and the proposed DWT-based features, as well as the adaptive matched filter (AMF) detector [2, 3], the constrained energy minimization (CEM) detector [2], the spectral angle mapper (SAM) algorithm [3], and the spectral information divergence (SID) algorithm [39, 40], are applied to the HSI data cubes, and the results are compared both visually and quantitatively using ROC curves and the corresponding AUROCs [42]. For more details on the AMF, CEM, SAM, and SID algorithms, the reader is encouraged to consult the corresponding references. The ROC curve depicts the true positive rate (TPR) as a function of the false positive rate (FPR). More specifically, thousands of different thresholds between the minimum and maximum values of the detector's output are picked. The class labels (background or target) for all pixels in the image are determined at each threshold. The FPR is calculated by dividing the number of false positives (background pixels predicted as target) by the total number of pixels in the image. The TPR is the ratio of the number of correct detections (target pixels predicted as target) to the total number of true target pixels. For the ROC curves, the results of the top four detection algorithms are shown: the AMF detector, the SID detector, the SFJTC technique using the original signatures, and the SFJTC technique using the optimal DWT coefficient combination (SFJTC_DWT).

Figures 14 and 15 show the ROC curves corresponding to the urban and vegetative CASI scenery with a 10 dB SNR. Note that the scale for the horizontal axes depicting the FPR ranges from 0 to 1 %. Tables 6 and 7 provide the AUROCs of detection for the traditional and SFJTC algorithms on the urban and vegetative CASI scenery. Similarly, Figures 16 and 17 show the ROC curves corresponding to the urban and vegetative HYDICE scenery with an 8 dB SNR, while Tables 8 and 9 provide the corresponding AUROCs.

As Table 6 shows, SFJTC_DWT provides a marginal improvement over the SFJTC technique using the original signatures. With the exception of *CASI_urban_15*, SFJTC_DWT provides the largest AUROC values. The ROC curves shed more light onto the performance of these detectors. Figure 14 clearly shows that

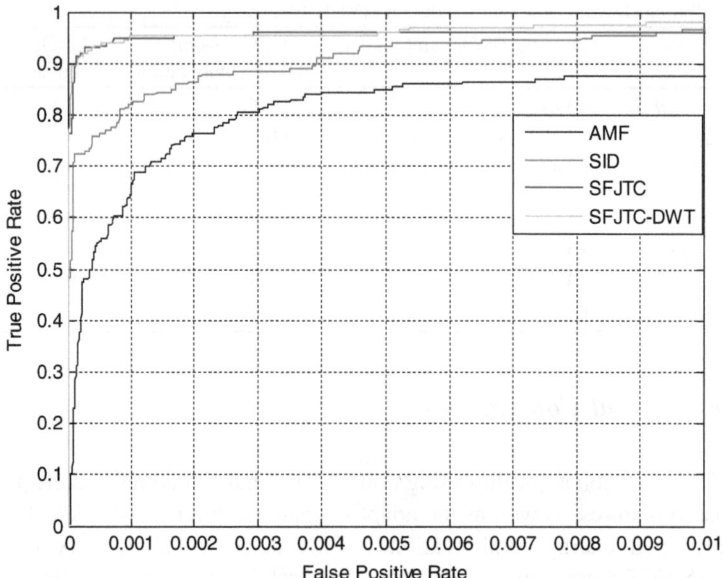

Fig. 14 SFJTC ROC curve comparisons for *CASI_urban_10*

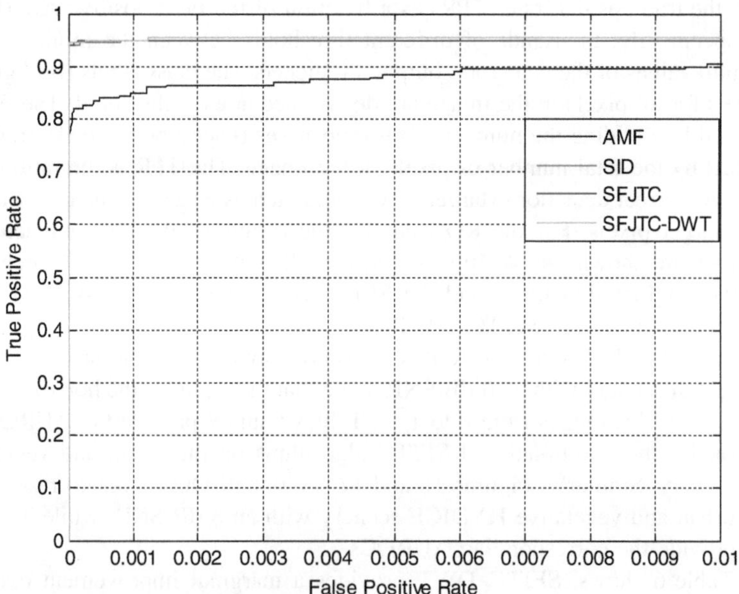

Fig. 15 SFJTC ROC curve comparisons for *CASI_veg_10*

Table 6 SFJTC AUROC comparisons for urban CASI scenery

Data Cube	AMF	CEM	SAM	SID	SFJTC	*SFJTC_DWT*
CASI_urban_15	0.9995	0.3889	0.2464	0.9951	0.9980	0.9987
CASI_urban_12	0.9923	0.4052	0.2518	0.9951	0.9980	0.9987
CASI_urban_10	0.9746	0.4151	0.2629	0.9947	0.9980	0.9987
CASI_urban_8	0.9013	0.4538	0.2558	0.9857	0.9967	0.9974

Table 7 SFJTC AUROC comparisons for vegetative CASI scenery

Data Cube	AMF	CEM	SAM	SID	SFJTC	*SFJTC_DWT*
CASI_veg_15	0.9998	0.0957	0.3483	0.9915	0.9916	0.9917
CASI_veg_12	0.9892	0.0969	0.3517	0.9915	0.9916	0.9882
CASI_veg_10	0.9662	0.1377	0.3538	0.9915	0.9916	0.9917
CASI_veg_8	0.9568	0.2204	0.3765	0.9915	0.9916	0.9917

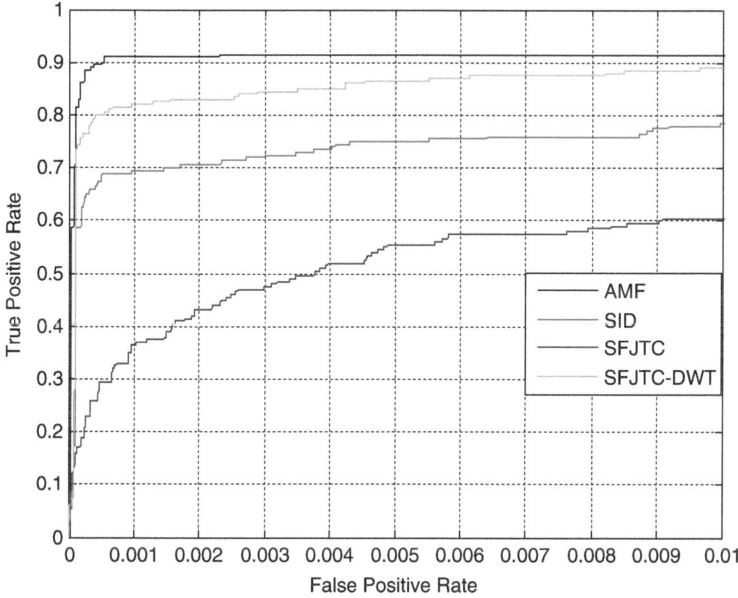

Fig. 16 SFJTC ROC curve comparisons for *HYDICE_urban_8*

both SFJTC techniques yield steeper ROC curves than the SID and AMF algorithms at FPRs below 1 %. For the vegetative CASI scenery, the AMF and SID algorithms provide the largest AUROCs for the two data cubes with least variability, while both SFJTC techniques provide the largest AUROCS for the two data cubes with the most spectral variability.

As Table 7 shows, the SID algorithm and SFJTC techniques provide the most consistent AUROCs on the vegetative CASI scenery.

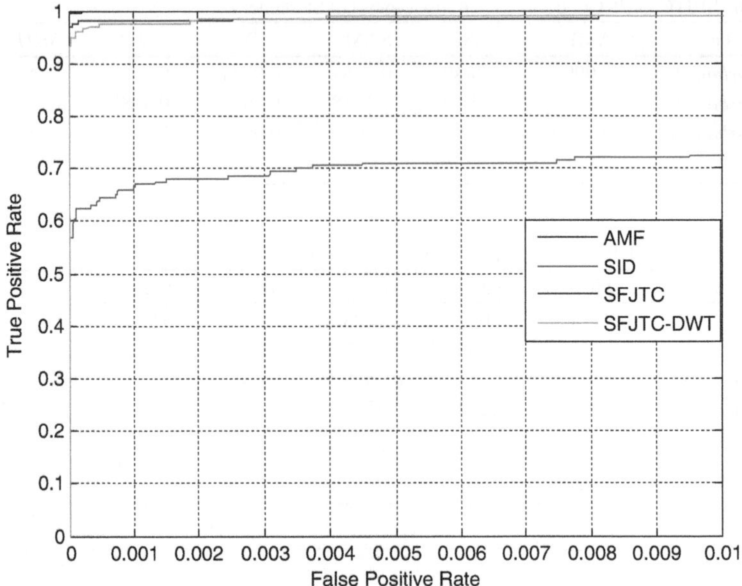

Fig. 17 SFJTC ROC curve comparisons for *HYDICE_veg_8*

Table 8 SFJTC AUROC comparisons for urban HYDICE scenery

Data Cube	AMF	CEM	SAM	SID	SFJTC	*SFJTC_DWT*
HYDICE_urban_15	1.0000	0.9656	0.9463	1.0000	1.0000	1.0000
HYDICE_urban_12	0.9950	0.9501	0.9459	1.0000	0.9999	1.0000
HYDICE_urban_10	0.9780	0.9342	0.9460	0.9999	0.9992	0.9999
HYDICE_urban_8	0.9434	0.9066	0.9457	0.9800	0.9715	0.9752

Table 9 SFJTC AUROC comparisons for vegetative HYDICE scenery

Data Cube	AMF	CEM	SAM	SID	SFJTC	*SFJTC_DWT*
HYDICE_veg_15	1.0000	1.0000	0.6948	1.0000	0.9986	1.0000
HYDICE_veg_12	1.0000	1.0000	0.6974	1.0000	0.9986	1.0000
HYDICE_veg_10	1.0000	1.0000	0.6913	0.9857	0.9986	1.0000
HYDICE_veg_8	1.0000	0.9992	0.6925	0.9370	0.9986	0.9995

As Table 8 shows, both SFJTC techniques and the SID algorithm yield stellar detection results for the first three data cubes of the urban HYDICE scenery. For *HYDICE_urban_8*, the data cube with the heaviest spectral variability, the SID provides a slightly larger AUROC than the SFJTC_DWT technique. Although the AMF and SFJTC_DWT algorithms have lower AUROCs than the SID for *HYDICE_urban_8*, Fig. 16 shows that they yield steeper ROC curves at FPRs below 1 %.

Table 10 Summary statistics of SFJTC AUROC comparisons

Scenery	AMF	CEM	SAM	SID	SFJTC	SFJTC_DWT
Urban CASI	0.9669 (0.0450)	0.4158 (0.0276)	0.2542 (0.0069)	0.9926 (0.0047)	0.9977 (0.0006)	0.9984 (0.0006)
Vegetative CASI	0.9780 (0.0199)	0.1376 (0.0585)	0.3576 (0.0128)	0.9915 (0)	0.9916 (0)	0.9908 (0.0018)
Urban HYDICE	0.9791 (0.0256)	0.9391 (0.0252)	0.9460 (0.0002)	0.9950 (0.0100)	0.9926 (0.0141)	0.9938 (0.0124)
Vegetative HYDICE	1.0000 (0)	0.9998 (0.0004)	0.6940 (0.0027)	0.9807 (0.0299)	0.9986 (0)	0.9999 (0.0002)
All	0.9810 (0.0277)	0.6231 (0.3738)	0.5630 (0.2836)	0.9900 (0.0153)	0.9951 (0.0071)	0.9957 (0.0067)

Both SFJTC techniques yield similar results for the vegetative HYDICE scenery, as shown in Table 9. The AMF algorithm yields perfect detection rates for all levels of spectral variability, while the CEM and SID algorithms yield slightly lower AUROCs. With the exception of the SAM algorithm, all of the algorithms yield superior results for the vegetative HYDICE scenery. For *HYDICE_veg_8*, the SID algorithm breaks down compared to the AMF and SFJTC algorithms as shown in Fig. 17 and Table 9.

Table 10 provides the first and second-order statistics (mean, standard deviation) of the AUROCs for the SFJTC techniques and traditional algorithms. For the urban CASI scenery, the most difficult scenery, SFJTC_DWT outperforms the other algorithms with the largest mean and smallest standard deviation AUROC values. For the vegetative CASI scenery, the SID and SFJTC algorithms provide the best results, with the SFJTC_DWT technique right behind them. For the urban HYDICE scenery, the SFJTC_DWT technique does second best, just shy of the SID algorithm. The AMF algorithm and SFJTC_DWT technique perform better than the SID and remaining algorithms on the vegetative HYDICE scenery. Over all of the scenes for both sensors, SFJTC_DWT is the best performer, providing the largest mean AUROC and smallest standard deviation of AUROC values.

5 Conclusion

We have addressed spectral variability in SFJTC-based target detection scenarios by exploring the use of the DWT coefficients of HS signatures as features for detection. We developed a supervised training algorithm that automatically determines an optimal combination of coefficients from a three-level DWT decomposition of the data. The training algorithm is simple yet adaptive, operating on the original target signature and randomly selected signatures from the input scene. Our experiments were conducted on real urban and vegetative HS scenery from both CASI and HYDICE sensors with scenarios of target spectral variability ranging from light to heavy. Results show that the use of the SFJTC technique with the optimal selected DWT coefficients provides increased or identical detection performance compared to using the original signatures. Furthermore, the results show that the proposed scheme provides larger mean AUROC values than current stochastic and deterministic detection algorithms [41].

Acknowledgments The authors wish to thank Drs. S. Ochilov, E. Sarigul and W. A. Sakla for many rewarding discussions.

References

1. Slater, D., Healey, G.: A spectral change space representation for invariant material tracking in hyperspectral images. Proc SPIE **3753**, 308–317 (1999)
2. Manolakis, D., Marden, D., Shaw, G.: Hyperspectral image processing for automatic target detection applications. Linc Lab J **14**, 79–114 (2003)
3. Manolakis, D., Shaw, G.: Detection algorithms for hyperspectral imaging applications. IEEE Signal Process. Mag. **19**, 29–43 (2002)
4. Mahalanobis, A., Muise, R.R., Stanfill, S.R.: Quadratic correlation filter design methodology for target detection and surveillance applications. Appl. Opt. **43**, 5198–5205 (2004)
5. Yamany, S.M., Farag, A.A., Hsu, S.-Y.: A fuzzy hyperspectral classifier for automatic target recognition (ATR) systems. Pattern Recogn. Lett. **20**, 1431–1438 (1999)
6. Manolakis, D.: Taxonomy of detection algorithms for hyperspectral imaging applications. Opt. Eng. **44**(6), 1–11 (2005)
7. Kay, S.M.: Fundamentals of statistical signal processing. Englewood Cliffs, New Jersey (1998)
8. Fisher, R.A.: Multiple measures in taxonomic problems. Ann. Eugenics **7**, 179–188 (1936)
9. Reed, I.S., Yu, X.: Adaptive multiple-band CFAR detection of an optical pattern with unknown spectral distribution. IEEE Trans Acoust. Speech, Signal Process. **38**, 1760–1770 (1990)
10. Center for the Study of Earth from Space (CSES), SIPS User's Guide, The spectral image processing system, vol. 1.1, pp. 74. University of Colorado, Boulder, (1992)
11. Weaver, C.S., Goodman, J.W.: A technique for optically convolving two functions. Appl. Opt. **5**, 1248–1249 (1966)
12. Yu, F.T.S., Ludman, J.E.: Microcomputer based programmable joint transform correlator for automatic pattern recognition and identification. Opt. Lett. **11**, 395–397 (1986)
13. Javidi, B., Tang, Q.: Chirp-encoded joint transform correlators with a single input plane. Appl. Opt. **33**, 227–230 (1994)
14. Alam, M.S., Goh S. F. Dacharaju, S.: Three-dimensional color pattern recognition using fringe-adjusted joint transform correlation with CIELab coordinates, accepted for publication, IEEE Trans. Instrum. Meas. **58**, 2176-2184 (2009)
15. Alam, M.S., Karim, M.A.: Fringe-adjusted joint transform correlation. Appl. Opt. **32**, 4344–4350 (1993)
16. Alam, M.S., Haque, M., Khan, J.F., Kettani, H.: Fringe-adjusted joint transform correlator based target detection and tracking in forward looking nfrared image sequence. Opt. Eng. **43**, 1407–1413 (2004)
17. Islam, M.N., Alam, M.S. Karim, M.A.: Pattern recognition in hyperspectral imagery using 1D shifted phase-encoded joint transform correlation. J. Opt. Commun. **281**, 4854–4861 (2008)
18. Alam, M.S., Ochilov, S.: Target detection in hyperspectral imagery using one-dimensional fringe-adjusted joint transform correlation. Proc. SPIE **6245**, 624505 (2006)
19. Alam, M.S., Bal, A., Horache, E.H., Goh, S.F., Loo, C.H., Regula, S.P., Sharma, A.: Metrics for evaluating the performance of joint-transform-correlation-based target recognition and tracking algorithms. Opt. Eng. **44**, 067005 (2005)
20. Wang, Q., Guo, Q., Zhou, J., Lin, Q.: Nonlinear joint fractional Fourier transform correlation for target detection in hyperspectral image. Opt. Laser Technol **44**, 1897–1904 (2012)
21. Jutamulia, S., Storti, G.M., Gregory, D.A., Kirsch, J.C.: Illumination-independent high-efficiency joint transform correlator. J. Appl. Opt. **30**, 4173–4175 (1991)
22. Alam, M.S., Ochilov, S.: Spectral fringe-adjusted joint transform correlation. Appl. Opt. **49**, B18–B25 (2010)
23. Alam, M.S., Karim, M.A.: Multiple target detection using a modified fringe-adjusted joint transform correlator. J. Opt. Eng. **33**, 1610–1617 (1994)

24. Sakla, W. Sakla, A., Alam, M.S.: Deterministic hyperspectral target detection using the DWT and spectral fringe-adjusted joint transform correlation (Invited Paper). In: Proceedings of the SPIE Conference on Automatic Target Recognition, vol. 6967, pp. 1–11 (2008)
25. DeVore, R.A., Jawerth, B., Lucier, B.J.: Image compression through wavelet transform coding. IEEE Trans. Inf. Theory **38**, 719–746 (1992)
26. Chang, S.G., Yu, B., Vetterli, M.: Adaptive wavelet thresholding for image denoising and compression. IEEE Trans. Image Process. **9**, 1532–1546 (2000)
27. Chang, T., Kuo, C.: Texture analysis and classification with tree-structured wavelet transform. IEEE Trans. Image Process. **2**, 429–441 (1993)
28. Bruce, L.M., Li, J.: Wavelets for computationally efficient hyperspectral derivative analysis. IEEE Trans. Geosci. Remote Sens. **39**, 1540–1546 (2001)
29. Bruce, L.M., Koger, C.H., Li, J.: Dimensionality reduction of hyperspectral data using discrete wavelet transform feature extraction. IEEE Trans. Geosci. Remote Sens. **40**, 2331–2338 (2002)
30. Kaewpijit, S., Le Moigne, J., El-Ghazawi, T.: Automatic reduction of hyperspectral imagery using wavelet spectral analysis. IEEE Trans. Geosci. Remote Sens. **41**, 863–871 (2003)
31. Bruce, L.M., Morgan, C., Larsen, S.: Automated detection of subpixel hyperspectral targets with continuous and discrete wavelet transforms. IEEE Trans. Geosci. Remote Sens. **39**, 2217–2226 (2001)
32. Mallat, S.: A theory for multiresolution signal decomposition: the wavelet representation. IEEE Trans. Pattern Anal. Mach. Intell. **11**, 674–693 (1989)
33. Vetterli, M., Kovacevic, J.: Wavelets and Subband Coding. Prentice Hall, Upper Saddle River (1995)
34. Mallat, S.: A Wavelet Tour of Signal Processing, 2nd edn. Academic Press, New York (1999)
35. ITRES Research http://www.itres.com, accessed in 2007.
36. Schowengerdt, R.A.: Remote Sensing, 2nd edn. Academic Press, San Diego (1997)
37. Chein, I.C., Heinz, D.C.: Constrained subpixel target detection for remotely sensed imagery. IEEE Trans. Geosci. Remote Sens. **38**(3), 1144–1159 (2000)
38. Jain, A.K.: Fundamentals of Digital Image Processing. Prentice-Hall, New Jersey (1989)
39. Chang, C.-I.: Hyperspectral Imaging: techniques for Spectral Detection and Classification. Kluwer Academic, New York (2003)
40. Chang, C.-I.: An information-theoretic approach to spectral variability, similarity, and discrimination for hyperspectral image analysis. IEEE Trans. Inf. Theory **46**, 1927–1932 (2000)
41. Sakla, A., Sakla, W., Alam, M.S.: Hyperspectral target detection via discrete wavelet-based spectral fringe-adjusted joint transform correlation. Appl. Opt. **50**, 5545–5554 (2011)
42. Parker, D.R., Gustafson, S.G., Ross, T.D.: Receiver operating characteristic and confidence error metrics for assessing the performance of automatic target recognition systems. Opt. Eng. **44**, 097202 (2005)

An Overview of Distributed Tracking and Control in Camera Networks

A. T. Kamal, C. Ding, A. A. Morye, J. A. Farrell and Amit K. Roy-Chowdhury

Abstract In many applications, the information from all the cameras are transmitted to a centralized processing unit. However, for larger networks, centralized schemes can be unscalable and they also exhibit the risk of having a single point of failure. These are some of the many reasons why distributed schemes are being chosen over centralized schemes in various applications nowadays. This chapter provides a comprehensive overview of state-of-the-art distributed tracking and control strategies using a camera network and identifies the main directions of future work.

1 Introduction

Networks of video cameras are being installed in many applications, like surveillance and security, disaster response, and environmental monitoring, among others. Currently, most of the data collected by such networks is analyzed manually, a task that is extremely tedious and reduces the potential of the installed networks. Therefore, it is essential to develop tools for automatically analyzing the data collected from these cameras and summarizing the results in a manner that is meaningful to the end user. This motivates the study of camera networks as a challenging research problem and a pressing social necessity.

A. T. Kamal · J. A. Farrell · A. K. Roy-Chowdhury (✉)
Department of Electrical Engineering, University of California,
Riverside, CA 92521, USA
e-mail: amitrc@ee.ucr.edu

C. Ding · A. A. Morye
Department of Computer Science and Engineering, University of California,
Riverside, CA 92521, USA

Augment Vis Real (2014) 6: 207–234
DOI: 10.1007/8612_2012_10

Published Online: 20 February 2013

In many applications, it is desirable that the video analysis tasks be distributed over the network. There may not be enough bandwidth and transmission power available to send all the data to a central station. Furthermore, security of the transmission and interception by a hostile opponent may be a factor in some applications. A centralized scheme necessitates setting up a processing unit with significantly different capabilities than those of the individual cameras. This may be challenging in many field operations, like disaster management or operations in hostile environments. Also, the centralized node introduces a single point of failure risk. Finally, even if it is possible to securely transmit all the data to a central unit, it may not be possible to efficiently analyze all the data at a single central node given the complex nature of these systems and the environments where they are deployed.

For distributed video analysis, the cameras would act as autonomous agents making decisions in a decentralized manner. However, the individual decisions of the cameras need to be coordinated so that there is consensus about the task (e.g., tracking, camera parameter assignment, activity recognition) even if each camera is an autonomous agent. Thus, the cameras need to analyze the raw data locally, exchange only distilled information that is relevant to the collaboration, and reach a shared, global analysis of the scene. Research in the area of distributed camera networks is very much in its early years and is a promising future direction, especially in the application domain of multi-agent systems equipped with video sensors.

We envision a system where each of the cameras will have its own embedded processing unit capable of basic video processing. This chapter provides a comprehensive overview of state-of-the-art distributed tracking and control strategies using a camera network and identifies the main directions of future work. Through calibration, the geometrical relationships among the cameras can be constructed. The cameras also need to have communication capabilities so that they can share information with each other.

Our proposed information processing and control structure is shown in Fig. 1. Each of the cameras in our network has its own embedded target detection module, a distributed tracker that provides an estimate on the state of each target in the scene, and finally a distributed camera control mechanism. Given the target detection module, this chapter provides a distributed solution for the tracking and control problems. Targets will be tracked using measurements from multiple cameras that may not have direct communication between each other. Neighboring cameras communicate with each other to come to a consensus about the estimated position of each target. Similarly, camera control parameters will be assigned based on information exchanged with neighboring cameras within a cooperative control framework.

We start by reviewing the recent works in the distributed estimation and control literature. Next, we review the Kalman Consensus Filter (KCF) [1] which is a well known consensus-based distributed estimation framework. Then we show the issues with using KCF in a camera-network and present the Generalized Kalman Consensus Filter (GKCF) [2] algorithm which can deal with most of the issues present with KCF when applied in a camera network. Next, we present the

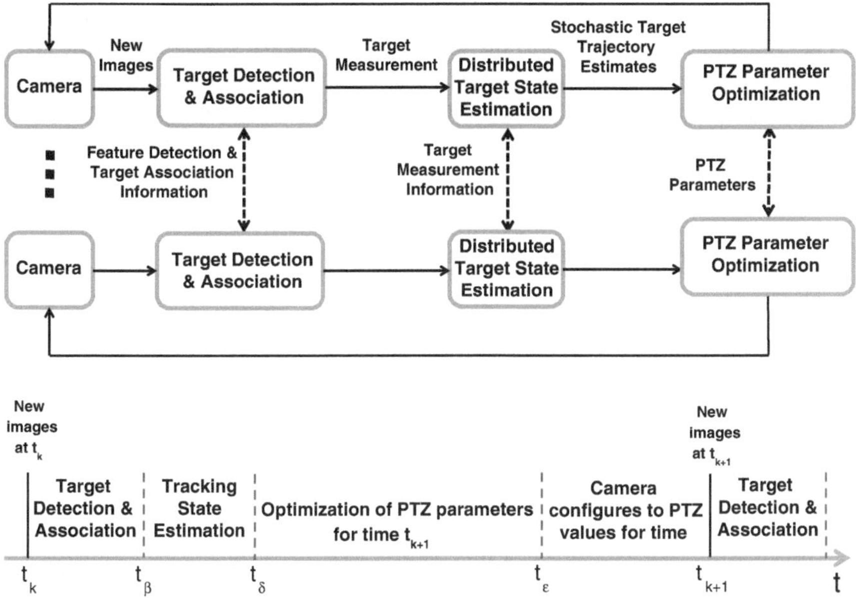

Fig. 1 Overall system diagram depicting our proposed framework for collaborative sensing in an active, distributive camera network. The user criteria can define what performance metrics the network will optimize. The user criteria could include covering the entire area at a desired resolution, obtaining facial shots, maximizing image resolution and so on. If the user does not specify any additional criteria, the tracking error will be minimized

Information-Weighted Consensus Filter (ICF) [3] where the distributed estimate meets desired performance guarantees. In the second half of this chapter we present an approach to optimize various scene analysis performance criteria through distributed control of a dynamic camera network [4]. A camera control framework using a distributed optimization approach allows selection of parameters to acquire images that best optimize the user-specified requirements. The uncertainty in state estimation when deciding upon camera settings is also considered in the framework. Experimental results demonstrate the real-life performance of the method.

1.1 Related Work

There have been a few papers in the recent past that deal with networks of video sensors. A review of current camera network research is available in [5]. Particular interest has been focused on learning a network topology [6, 7], i.e., configuring connections between cameras and entry/exit points in their view. Some of the existing methods on tracking over the network, include [8–10]. Other interesting problems in camera networks, like object/behavior detection and matching across cameras, camera handoff and camera network configuration have been addressed in [11–16]. In [17], a solution to the problem of optimal camera placement given some coverage constraints was presented and can be used to come up with an initial camera configuration. There has also been recent work on tracking people in a multi-camera setup [18, 19]. However, these methods did not address the issue of distributed processing.

1.1.1 Distributed Estimation in Camera Networks

Recently, some problems in distributed video analysis have been addressed in multi-camera networks. An algorithm for the calibration of a distributed camera network was presented in [20]. The problems of object detection, tracking, recognition and pose estimation in distributed camera network were considered in [21, 22]. These methods were designed for a network composed of static cameras; the dynamics of active camera networks were not taken into account.

Among many types of distributed estimation schemes, *consensus algorithms* [23] have recently gained immense attention for their scalability, robustness and ease of implementation in various scenarios. Consensus algorithms are protocols that are run individually by each agent where each agent communicates with just its network neighbors and corrects its own information iteratively using the information sent by its neighbors. The protocol, over multiple iterations, ensures the convergence of all the agents in the network to a single consensus.

Consensus algorithms have been extended to perform various tasks in a network of agents such as various linear algebraic operations [24, 25], distributed state and parameter estimation frameworks such as the distributed maximum likelihood estimator (DMLE) [26] and distributed tracking such as [1–3]. A detailed review of distributed state estimation methods and comparisons with centralized and decentralized approaches can be found in [27]. These distributed state and parameter estimation frameworks have been applied in various fields including camera networks for distributed implementations of 3-D point triangulation, pose estimation [24], tracking [28], action recognition [28, 29], and collaborative tracking and camera control [30].

Most existing distributed estimation methods assume that each target can be viewed by each sensor which may not be true for many application scenarios, especially for a camera network (see Fig. 2) where each camera can view only a limited portion of the entire area. This limits the observability of each sensor to a subset of all the targets. Later in this chapter, we show how to design a distributed multi-target tracking scheme which is suited for such sensors with limited field-of-view (FOV).

1.1.2 Distributed Control in Camera Networks

An overview of some main video processing techniques and currents trends for video analysis in pan-tilt-zoom (PTZ) camera networks can be found in [31].

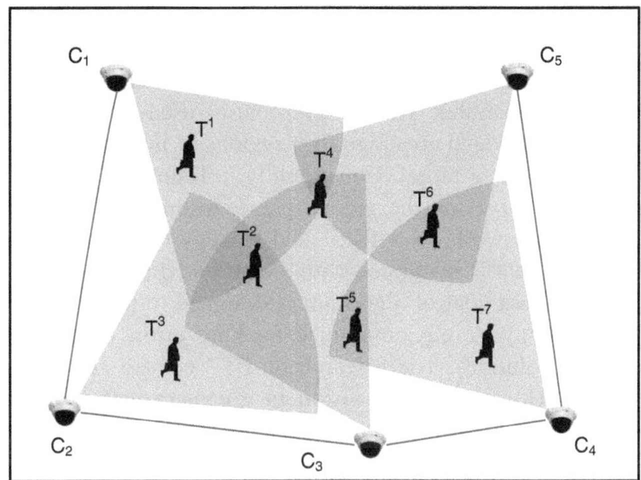

Fig. 2 In this figure, there are five sensing nodes, $C_1, C_2, \ldots, C_5$ and seven targets $T^1, T^2, \ldots T^7$. The *solid blue lines* show the communication channels between different nodes. This figure also depicts the presence of "naive" nodes. For example, C_1 gets direct measurements about T_1 which it shares with its immediate network neighbor, C_2. However, the rest of the cameras, i.e., C_3, C_4, C_5 do not have *direct* access to measurements of T^1 and thus are naive w.r.t. T^1's state

Cooperation in a network consisting of static and PT cameras was considered in [32]. The parameters of PT cameras were determined by centering the desired targets in the field of view and the cooperation is only between the static camera set and the PT camera set, i.e., the static cameras check if a PT camera is tracking a correct target. A related work that deals with the control of a camera network with PTZ cameras is [33]. Here, a virtual camera network environment was used to demonstrate a camera control scheme that is a mixture between a distributed and a centralized scheme using both passive and active PTZ cameras. Their work focused on how to group cameras which are relevant to the same task in a centralized manner while maintaining the individual groups decentralized. In [4], the authors consider a completely distributed solution using a game-theoretic framework for camera parameter control and implicit target assignment. A game-theoretic framework for vehicle-to-target assignment was proposed in [34] in the context of distributed sensor networks. However, in that work the targets were stationary and each vehicle was assigned to one target. That work did not consider the constraints imposed by video cameras as directional sensors. Moreover, it did not consider that each camera can observe multiple targets and multiple cameras can observe each target (many-to-many mapping). A camera handoff approach using game theory was presented in [14]. That method, however, considered only a set of static cameras and does not deal with the problem of persistently observing targets with varying resolutions over a large geographic area using a dynamic camera network with overlapping and non-overlapping FOVs.

A game-theoretic approach to camera control was presented in [35] but limited only to the area coverage problem. This was expanded on in [36] to a distributed tracking and control approach. It required the camera control and tracking to run independently and in parallel. The camera control used game theory to assign camera settings that provided coverage over regions of interest while maintaining a high resolution shot of a target. Concurrently, a Kalman-Consensus filter provided tracks of each target on the ground plane. In this chapter, we provide an overview of our recent work on a distributed strategy for controlling the parameters of the cameras that is integrated with a distributed tracking algorithm [2–4, 26, 37, 38]. The camera control algorithm is designed to maximize performance criteria for scene analysis (e.g., minimize tracking error, obtain a facial shot, maximize image resolution). The research on PTZ camera control is related to active vision [39]. However active vision in a camera network is a relatively unexplored area that would involve cooperation and coordination between many cameras. Coordinated sensing and tracking strategies have also been applied to mobile camera platforms [37, 40–44].

2 Distributed Estimation for Tracking

2.1 Problem Formulation

Consider a sensor network with N_C sensors. The communication in the network can be represented using an undirected connected graph $\mathcal{G} = (\mathcal{C}, \mathcal{E})$. The set $\mathcal{C} = \{C_1, C_2, \ldots, C_{N_C}\}$ contains the vertices of the graph and represents the sensor nodes. The set $\mathcal{E}$ contains the edges of the graph which represents the available communication channels between different nodes. The set of nodes that each have a direct communication channel with node C_i (sharing an edge with C_i) is represented by $\mathcal{N}_i$.

The true state of a target is represented by $\mathbf{x}(t) \in \mathcal{R}^p$. In this description of the distributed state estimation problem, we assume that the data association is given. For simplicity of notation, time index t will be dropped where the issue under consideration can be understood without it. Each node has a prior estimate of $\mathbf{x}$ as $\bar{\mathbf{x}}_i^- \in \mathcal{R}^p$. The error in the prior estimate at C_i is $\boldsymbol{\eta}_i = \bar{\mathbf{x}}_i^- - \mathbf{x} \in \mathcal{R}^p$ with covariance $\mathbf{P}_i^- \in \mathcal{R}^{p \times p}$. The prior information/precision matrix of node C_i is denoted as $\mathbf{J}_i^- \in \mathcal{R}^{p \times p}$, where

$$\mathbf{J}_i^- = (\mathbf{P}_i^-)^{-1}. \tag{1}$$

This information matrix is useful in the distributed estimation task.

The observation of node C_i is denoted by $\mathbf{z}_i \in \mathcal{R}^{m_i}$ with noise covariance $\mathbf{R}_i \in \mathcal{R}^{m_i \times m_i}$, where m_i is the length of the measurement vector at node C_i. The observations from all the nodes are modeled as

$$\mathcal{Z} = \mathcal{H}\mathbf{x} + \mathbf{v}. \tag{2}$$

Here, $\mathcal{Z} = [\mathbf{z}_1^T, \mathbf{z}_2^T, \ldots, \mathbf{z}_N^T]^T \in \mathcal{R}^m$ and observation matrix $\mathcal{H} = [\mathbf{H}_1^T, \mathbf{H}_2^T, \ldots, \mathbf{H}_N^T]^T \in \mathcal{R}^{m \times p}$ where, $\mathbf{H}_i \in \mathcal{R}^{m_i \times p}$ and $m = \sum_{i=1}^N m_i$. Observation noise $\mathbf{v}$ is assumed to be Gaussian with $\mathbf{v} \sim \mathcal{N}(\mathbf{0}, \mathcal{R}) \in \mathcal{R}^m$. The inverse of $\mathbf{R} \in \mathcal{R}^{m \times m}$ is denoted by $\mathbf{B} \in \mathcal{R}^{m \times m}$. The measurements are assumed to be uncorrelated across nodes. Thus, the measurement information matrix is block diagonal and can be expressed as,

$$\mathcal{B} = \begin{bmatrix} \mathbf{B}_1 & \mathbf{0} & \ldots & \mathbf{0} \\ \mathbf{0} & \mathbf{B}_2 & & \vdots \\ \vdots & & \ddots & \\ \mathbf{0} & \ldots & & \mathbf{B}_N \end{bmatrix}. \tag{3}$$

Here, $\mathbf{B}_i = \mathbf{R}_i^{-1} \in \mathcal{R}^{m_i \times m_i}$.

The state evolution is modeled using the following linear dynamical model,

$$\mathbf{x}(t+1) = \mathbf{\Phi}\mathbf{x}(t) + \gamma(t). \tag{4}$$

Here $\mathbf{\Phi}$ is the state propagation matrix and process noise $\gamma(t) \sim \mathcal{N}(\mathbf{0}, \mathbf{Q})$.

2.2 Average Consensus

Average consensus [23] is a popular distributed algorithm to compute the arithmetic mean of some values $\{a_i\}_{i=1}^{N_C}$. Suppose, in a network of N_C nodes, each node i has an estimate a_i. The goal is to compute the average value of these quantities, i.e., $\frac{1}{N_C}\sum_{i=1}^N a_i$, in a distributed manner.

In average consensus algorithm, each node initializes its initial consensus state as $a_i(0) = a_i$ and then runs the following protocol iteratively:

$$a_i(k) = a_i(k-1) + \epsilon \sum_{j \in \mathcal{N}_i} \big(a_j(k-1) - a_i(k-1)\big). \tag{5}$$

At the beginning of iteration k, a node C_i sends its previous estimate $a_i(k-1)$ to its immediate network neighbors $C_j \in \mathcal{N}_i$ and also receives the neighbors' previous estimates $a_j(k-1)$. Then it updates its estimate using (5). By iteratively doing so, the values of the estimate at all the nodes converge to the average of the initial values. The average consensus algorithm can be used to compute the average of vectors and matrices by applying it to their individual elements separately. Note that average consensus treats all nodes as having equal and uncorrelated information about the quantity a.

The rate parameter ϵ should be chosen between 0 and $\frac{1}{\Delta_{max}}$, where Δ_{max} is the maximum degree of the network graph $\mathcal{G}$. Choosing larger values of ϵ will result in faster convergence, but choosing values equal or more than Δ_{max} will render the algorithm unstable. More information about average consensus and about the rate parameter ϵ can be found in [23]. The average consensus algorithm has been extended to situations with varying network topology [45, 46] and imperfect communication links (where packet loss may occur) [47].

In this article, the average consensus algorithm will be utilized to compute certain averages, one of which is the average of information matrices at different nodes. As, information matrices must be symmetric positive semi-definite, we must ensure that applying average consensus on information matrices preserves this property. To check this fact, we can rearrange (5) as

$$a_i(k) = (1 - \Delta_i\epsilon)a_i(k-1) + \sum_{j \in \mathcal{N}_i} \epsilon a_j(k-1). \tag{6}$$

Here, Δ_i is the degree of the node C_i in the communication graph. The coefficients ϵ and $(1 - \Delta_i\epsilon)$ are always positive. A linear combination (with positive weights) of symmetric positive semi-definite matrices is also a symmetric positive semi-definite matrix. Thus, average consensus on information matrices preserves their symmetric positive semi-definite property.

2.3 Kalman Consens Filter for Distributed Tracking

We briefly review the Kalman Consensus Filter (KCF) [1] (see Algorithm 1), analyze its applicability to distributed tracking, and propose alternatives that take into account the characteristics of visual sensors.

Algorithm 1 KCF at C_i at time step t

Given $\mathbf{J}_i^-(t)$, $\bar{\mathbf{x}}_i^-(t)$, $\mathbf{H}_i$, ϵ and K

(1) Get measurement $\mathbf{z}_i$ and measurement information matrix $\mathbf{B}_i$

(2) Compute measurement information vector and matrix

$$\mathbf{u}_i = \mathbf{H}_i^T \mathbf{B}_i \mathbf{z}_i \tag{7}$$

$$\mathbf{U}_i = \mathbf{H}_i^T \mathbf{B}_i \mathbf{H}_i \tag{8}$$

(3) Receive $\mathbf{u}_{i'}, \mathbf{U}_{i'}, \bar{\mathbf{x}}_{i'}^-$ from neighbors $i' \in \mathcal{N}_i$

(4) Fuse the information vectors and matrices and calculate weight matrices

$$\mathbf{y}_i = \sum_{i' \in \mathcal{N}_i \cup \{i\}} \mathbf{u}_{i'} \tag{9}$$

$$\mathbf{S}_i = \sum_{i' \in \mathcal{N}_i \cup \{i\}} \mathbf{U}_{i'} \tag{10}$$

(5) Compute Kalman Consensus estimate

$$\mathbf{M}_i = (\mathbf{J}_i^-(t) + \mathbf{S}_i)^{-1} \tag{11}$$

$$\gamma = \epsilon / (1 + ||(\mathbf{J}_i^-(t))^{-1}||), \quad ||X|| = tr(X^T X)^{\frac{1}{2}} \tag{12}$$

$$\bar{\mathbf{x}}_i^+(t) = \bar{\mathbf{x}}_i^-(t) + \mathbf{M}_i(\mathbf{y}_i - \mathbf{S}_i \bar{\mathbf{x}}_i^-(t)) + \gamma(\mathbf{J}_i^-(t))^{-1} \sum_{j \in \mathcal{N}_i} (\bar{\mathbf{x}}_j^-(t) - \bar{\mathbf{x}}_i^-(t)) \tag{13}$$

(6) Update the state of the Kalman-Consensus filter

$$\mathbf{J}_i^-(t+1) \leftarrow (\Phi \mathbf{M}_i \Phi^T + \mathbf{Q})^{-1} \tag{14}$$

$$\bar{\mathbf{x}}_i^-(t+1) \leftarrow \Phi \bar{\mathbf{x}}_i^+(t) \tag{15}$$

Here, K is the total number of imaging time instants. The KCF algorithm performs reasonably well when all the nodes in the network can directly observe the target. However, in the scenario where nodes have limited FOV, some nodes do not have direct measurement of the target and in these cases the performance of KCF may be highly degraded (as shown in Fig. 3). The issue of limited sensing range in the distributed estimation process has been considered in [48], where the authors considered the case where not all sensors get measurements of the target. However, the solution was not fully distributed; rather it was a hybrid solution consisting of a distributed and a centralized scheme for information fusion. The nodes used the KCF algorithm to update their state estimates. These state estimates were sent along with the state covariance information to a *fusion center*. In this chapter, we study the problem using a completely distributed architecture.

2.4 Limitations of KCF in Camera Networks

In a sensor network, even when the network has sufficient measurements for the state to be observable by a centralized processor, it may be the case that some nodes have limited observability, defined as the node not having sufficient information in its *local neighborhood* (consisting of the node and its immediate network neighbors) for the state to be observable. Due to limited observability and limited number of iterations, such nodes become *naive* about the target's state. A naive node contains less information about the state. If a naive node's estimate is given an equal weight in the information fusion scheme (as in KCF), the performance of the overall state estimation framework may decrease. The effect of naivety is severe in sparse networks where the total number of edges is much smaller than the maximum possible number of edges. Due to the presence of naive nodes the performance of the KCF might deteriorate. Next, we discuss various specific conditions that require attention for distributed estimation applied to sparse (e.g., camera) networks with naive nodes, and we propose solution strategies for each of them.

2.4.1 Average Versus Weighted Average

The basic KCF algorithm uses *average* consensus to combine state estimates from neighboring nodes (see Eq. (13)). With average consensus, the state estimates of all the nodes get the same weight in the summation. Since naive nodes do not have observations of the target, their estimates are often highly erroneous. This results in reduced performance in the presence of naive nodes.

2.4.2 Covariance/Information Matrix Propagation

The information matrix measurement update of Eq. (11) considers the node's own information matrix and the local neighborhood's measurement covariance. It does not account for cross covariance between the estimates by the node and its neighbors. In the theoretical proof of optimality for KCF, the cross covariances terms between neighbors' state estimates were present [1]. It has been stated in [1] that dropping these cross covariance terms is a valid approximation when the state estimate error covariance matrices are almost equal in all the nodes.

However, when C_i is naive w.r.t. T^j, $\mathbf{y}_i$ and $\mathbf{S}_i$ are both zero. Therefore, $\mathbf{M}_i = (\mathbf{J}_i^-)^{-1}$ at Eq. (11). Consequently, from Eq. (14) it can be seen that the diagonal elements of $\mathbf{J}_i^-$ tend to zero at each time update as long as C_i remains naive with respect to T^j. This makes the covariance matrix diverge. From this, it can be clearly seen that omitting the cross covariances in the covariance update equation is not valid for sparse networks with naive agents. The correlation between the two dependent variables is the unknown parameter making this computation difficult. There has been some work, e.g. [49] and [50], where the authors incorporated cross covariance information, which should lead to the optimum result. But, no method for computing these terms were provided and predefined fixed values were used instead.

2.4.3 Appropriate weighting of information

The measurement update term and consensus term in Eq. (13) are both functions of the prior state estimate $\bar{\mathbf{x}}_i^-(k)$. Both terms apply corrections to the prior state estimate, from different information sources. However, the relative weighting between these two innovation terms is not appropriate. This is usually not a big issue in sensor networks without naive nodes because every node's state estimate will be close to the consensus. In sparse networks, the estimates of naive nodes may lag behind by a significant time. This happens because naive nodes do not have direct access to new observation of a target, the only way they can get updated information about a target is through a neighbor's state estimate which was updated in the previous iteration. Thus a naive node might be multiple iterations away from getting new information about a target. This information imbalance can cause large oscillations. In the KCF algorithms this effect can be decreased by choosing a smaller rate parameter ϵ. However, decreasing ϵ yields slower convergence of the naive node's state estimate.

The above issues can be problematic for tracking applications involving a camera network with naive nodes. A naive node may associate an observation to a wrong target. This can affect the tracking performance of nodes that are actually observing the target by influencing them to drift away from their estimates. Since KCF is a very appropriate framework to build a distributed tracker in a camera network, the following two sections discuss algorithms proposed to address the above issues.

2.5 Generalized Kalman Consensus Filter

The Generalized Kalman Consensus Filter (GKCF) has the following characteristics.

(1) The consensus portion of the GKCF correction step at each node will take into account the state covariances of neighbors. The nodes will then converge towards the weighted mean, instead the unweighted mean.
(2) Each node and its neighbors' state covariance matrices will be used jointly at consensus step to update that node's error covariance matrix. This will prevent the state covariance of the naive nodes from monotonically increasing.
(3) Weighted average consensus will correct the prior estimate towards the weighted mean. Then the measurements in the local neighborhood will be used to update this consensus state and covariance, thus preventing the overcorrection issue mentioned above.

The GKCF algorithm is given in Algorithm 2. In the following, we will discuss how the different issues of using KCF in a camera-network are resolved in GKCF.

2.5.1 Weighted Average Consensus

Let the initial state estimate of each agent be $\bar{\mathbf{x}}_i^-$ with information matrix $\mathbf{J}_i^-$. As we use this information matrix term as weights in the weighted average consensus algorithm, the terms *weight* and *information matrix* will be used interchangeably. So, the global weighted average of the initial states is

$$\mathbf{x}^* = \left(\sum_{i=1}^{N_C} \mathbf{J}_i^-\right)^{-1} \sum_{i=1}^{N_C} \mathbf{J}_i^- \bar{\mathbf{x}}_i^- . \tag{16}$$

Define the weighted initial state of each agent as

$$\tilde{\mathbf{x}}_i = \mathbf{J}_i^- \bar{\mathbf{x}}_i^- . \tag{17}$$

Weighted average consensus [23] states that if the iterative update in Eqs. (24) and (25) is performed for all $i = 1, \ldots, N_c$, then each of the terms $\mathbf{W}_i(k)^{-1}\tilde{\mathbf{x}}_i(k)$ in Algorithm 2 tends to the global weighted average $\mathbf{x}^*$ as $k \to \infty$. As a by-product, the weights also converge to the average of the initial weights. Both these properties of the weighted average consensus are utilized in our approach.

Algorithm 2 GKCF at C_i at time step t

Given $\mathbf{J}_i^-(t)$, $\bar{\mathbf{x}}_i^-(t)$, $\mathbf{H}_i$, ϵ and K, let,

$$\tilde{\mathbf{x}}_i(0) \leftarrow \mathbf{J}_i^-(t)\bar{\mathbf{x}}_i^-(t) \tag{18}$$

$$\mathbf{W}_i(0) \leftarrow \mathbf{J}_i^-(t) \tag{19}$$

(1) Get measurement $\mathbf{z}_i$ and measurement information matrix $\mathbf{B}_i$
(2) Compute measurement information vector and matrix

$$\mathbf{u}_i = \mathbf{H}_i^T \mathbf{B}_i \mathbf{z}_i \tag{20}$$

$$\mathbf{U}_i = \mathbf{H}_i^T \mathbf{B}_i \mathbf{H}_i \tag{21}$$

(3) Receive $\mathbf{u}_{i'}$ and $\mathbf{U}_{i'}$ from neighbors and fuse the measurement information matrices and vectors

$$\mathbf{y}_i = \sum_{i' \in \mathcal{N}_i \cup \{i\}} \mathbf{u}_{i'} \tag{22}$$

$$\mathbf{S}_i = \sum_{i' \in \mathcal{N}_i \cup \{i\}} \mathbf{U}_{i'} \tag{23}$$

(4) Perform consensus

for k = 1 to K **do**

(a) Receive $\tilde{\mathbf{x}}_{i'}(k-1)$ and $\mathbf{W}_{i'}(k-1)$ from neighbors $i' \in \mathcal{N}_i$
(b) Update weighted average consensus estimates

$$\tilde{\mathbf{x}}_i(k) = \tilde{\mathbf{x}}_i(k-1) + \epsilon \sum_{i' \in \mathcal{N}_i} \left(\tilde{\mathbf{x}}_{i'}(k-1) - \tilde{\mathbf{x}}_i(k-1)\right) \tag{24}$$

$$\mathbf{W}_i(k) = \mathbf{W}_i(k-1) + \epsilon \sum_{i' \in \mathcal{N}_i} (\mathbf{W}_{i'}(k-1) - \mathbf{W}_i(k-1)) \tag{25}$$

endfor

(5) Update Prior

$$\bar{\mathbf{x}}_i^- \leftarrow \mathbf{W}_i(K)^{-1} \tilde{\mathbf{x}}_i(K) \tag{26}$$

(6) Compute Kalman consensus estimate

$$\mathbf{J}_i^+(t) \leftarrow \mathbf{W}_i(K) + \mathbf{S}_i \tag{27}$$

$$\bar{\mathbf{x}}_i^+(t) = \bar{\mathbf{x}}_i^- + (\mathbf{J}_i^+(t))^{-1} (\mathbf{y}_i - \mathbf{S}_i \bar{\mathbf{x}}_i^-) \tag{28}$$

(7) Propagate weight and weighted state estimate

$$\mathbf{J}_i^-(t+1) \leftarrow (\Phi \mathbf{J}_i^+(t)^{-1} \Phi^T + \mathbf{Q})^{-1} \tag{29}$$

$$\bar{\mathbf{x}}_i^-(t+1) \leftarrow \Phi \bar{\mathbf{x}}_i^+(t) \tag{30}$$

2.5.2 Covariance/Information Matrix Propagation

After communicating with its neighbors and prior to using measurement information, the optimal state estimate at C_i is a linear combination of the information from C_i and its neighbors. Since these variables are not independent, optimal estimation would require knowledge of the cross correlation structure between each pair of these random variables. Since, it is usually quite difficult to compute this cross correlation, we need some other way to approximate the covariance or in this case the information matrix.

The update operation of the information matrix $\mathbf{J}_i^-$ in Eq. (25) can be used as an approximation of the information matrix due to the incoming information from the neighbors' states. A property of the weighted average consensus is that the weights also converge to the average of the weights as the state estimates converge towards the weighted average. Thus, this kind of covariance/weight propagation enables the weights to be updated accordingly when informative state estimates arrive at a naive node.

After computing the state and weight estimates with all the available information, we need to propagate the weight and state in time. One should note that instead of propagating the state estimate, we have to propagate the weighted state estimate as necessitated by the weighted average consensus equations. Thus the weight propagation equation takes the form of Eq. (29).

2.5.3 Two-Stage Update

To weight the measurement and prior information appropriately, we divide the estimation process in two stages. First, as mentioned above, C_i updates its state and information matrix using its neighbors' states and information matrices. Next, we further update our state and information matrix with current measurement information, which we explain below.

Consider that a node that has completed Step 3 in Algorithm 2. If it did not have any observation, then $\mathbf{z}_i$ and $\mathbf{B}_i$ were set to zero. Using the fused information vector and matrix and the updated prior weight and state estimate (from the weighted average consensus step of Eqs. (25) and (26)) appropriately in a standard Distributed Kalman Filter, we get the final state and weight estimate at time t.

2.6 Information Weighted Consensus Filter

In both the KCF and GKCF, the cross-covariances between the priors across different nodes are not incorporated in the estimation framework. As the consensus progresses, the errors in the information at each node become highly correlated with each other. Thus, to compute the optimal state estimate, the error cross-covariances cannot be neglected. However, it is difficult to compute the cross-covariance in a distributed framework. We note that in a consensus-based framework, the state estimates at different nodes achieve reasonable convergence over multiple iterations. At this point, each node contains almost identical/redundant information. This fact can be utilized to compute the optimal estimate in a distributed framework without explicitly computing the cross-covariances.

Motivated by this idea, an Information-weighted Consensus Filter (ICF) algorithm was proposed for distributed state and parameter estimation [3], which is guaranteed to converge to the optimal centralized estimates as the prior state estimates become equal at different nodes i.e., the total number of iterations approach to infinity at the previous time step. It is also shown experimentally that even with limited number of iterations, the proposed algorithm achieves near-optimal performance. The issue of naivety and optimality is handled by proper information weighting of the prior and measurement information. The communication bandwidth requirement is also low for ICF.

2.6.1 Overview of ICF

The task in a centralized a posteriori estimation process is to estimate the state $\mathbf{x}$ from the measurements $\mathcal{Z}$ and prior state $\hat{\mathbf{x}}_c^-$ (with information matrix $\mathbf{J}_c^-$). In the centralized case, after the estimation at time $t-1$, we have the state estimate

$\hat{\mathbf{x}}_c^+(t-1)$ that is used as a prior ($\hat{\mathbf{x}}_c^-(t)$) for the estimation at time t. For the distributed case, each node will have its own prior $\bar{\mathbf{x}}_i^-(t)$. Ideally, for all i, $\bar{\mathbf{x}}_i^-(t)$ should be equal to $\hat{\mathbf{x}}_c^-(t)$. However, in practice, due to limited number of consensus iterations at previous time steps, there may be some discrepancies among the priors in the distributed case. As the number of iterations increase or with more time steps, the state estimates at each node becomes very similar. The ICF algorithm utilizes this fact in its framework.

The ICF algorithm is summarized in Algorithm 3. At each time step, if $k \to \infty$, the ICF guarantees that the priors for the next time step at each node will be equal to the optimal centralized one. This in turn sets the optimality condition for the next time step.

In reality, reaching true convergence may not be possible due to limited number of consensus iterations. The number of iterations needed to reach a reasonable convergence depends on the network size and number and position of naive nodes in the network graph. In Sect. 3.5, we will show experimentally that in case of only one or a few iterations, ICF is robust to small discrepancies between the state estimates across the nodes and achieves near optimal performance.

Algorithm 3 ICF at node C_i at time step t

Input: Prior state estimate $\bar{\mathbf{x}}_i^-(t)$, prior information matrix $\mathbf{J}_i^-(t)$, observation matrix $\mathbf{H}_i$, consensus rate parameter ϵ, total consensus iterations K, state transition matrix $\mathbf{\Phi}$, process covariance $\mathbf{Q}$ and total number of nodes N_C.

(1) Get measurement $\mathbf{z}_i$ and measurement information matrix $\mathbf{B}_i$

(2) Compute initial information matrix and vector

$$\mathbf{V}_i^0 \leftarrow \frac{1}{N_C}\mathbf{J}_i^-(t) + \mathbf{H}_i^T\mathbf{B}_i\mathbf{H}_i \tag{31}$$

$$\mathbf{v}_i^0 \leftarrow \frac{1}{N_C}\mathbf{J}_i^-(t)\bar{\mathbf{x}}_i^-(t) + \mathbf{H}_i^T\mathbf{B}_i\mathbf{z}_i \tag{32}$$

(3) Perform average consensus on $\mathbf{V}_i^0$ and $\mathbf{v}_i^0$ independently
for k = 1 to K **do**

(a) Send $\mathbf{V}_i^{k-1}$ and $\mathbf{v}_i^{k-1}$ to all neighbors $j \in \mathcal{N}_i$

(b) Receive $\mathbf{V}_j^{k-1}$ and $\mathbf{v}_j^{k-1}$ from all neighbors $j \in \mathcal{N}_i$

(c) Update:

$$\mathbf{V}_i^k \leftarrow \mathbf{V}_i^{k-1} + \epsilon \sum_{j \in \mathcal{N}_i}\left(\mathbf{V}_j^{k-1} - \mathbf{V}_i^{k-1}\right) \tag{33}$$

$$\mathbf{v}_i^k \leftarrow \mathbf{v}_i^{k-1} + \epsilon \sum_{j \in \mathcal{N}_i} \left(\mathbf{v}_j^{k-1} - \mathbf{v}_i^{k-1} \right) \tag{34}$$

end for

(4) Compute a posteriori state estimate andinformation matrix for time t

$$\mathbf{x}_i^+(t) \leftarrow (\mathbf{V}_i^K)^{-1} \mathbf{v}_i^K \tag{35}$$

$$\mathbf{J}_i^+(t) \leftarrow N_C \mathbf{V}_i^K \tag{36}$$

(5) Predict for next time step $(t + 1)$

$$\mathbf{J}_i^-(t+1) \leftarrow \left(\mathbf{\Phi}(\mathbf{J}_i^+(t))^{-1} \mathbf{\Phi}^T + \mathbf{Q} \right)^{-1} \tag{37}$$

$$\bar{\mathbf{x}}_i^-(t+1) \leftarrow \mathbf{\Phi} \bar{\mathbf{x}}_i^+(t). \tag{38}$$

Output: State estimate $\bar{\mathbf{x}}_i^+(t)$and information matrix $\mathbf{J}_i^+(t)$.

2.7 Comparative Evaluation of Distributed Estimation Approaches

In this section, we evaluate the performance of the proposed ICF algorithm in a simulated environment and compare it with other methods: the Centralized Kalman Filter (CKF) [51], the Kalman Consensus Filter (KCF) [1] and the Generalized Kalman Consensus Filter (GKCF) [2].

We simulate a camera network in this experiment, where the state estimation algorithms are used for tracking a target roaming within a 500×500 space. The target's initial state vector is random. The target's state vector is a $4D$ vector, with the $2D$ position and $2D$ velocity components. The initial speed is uniformly picked from 10–20 units per time step, with a random direction uniformly chosen from 0 to 2π. The targets evolve for 40 time steps using the target dynamical model of Eq. (4). The state transition matrix $\mathbf{\Phi}$ and process covariance $\mathbf{Q}$ are chosen as the following

$$\mathbf{\Phi} = \begin{bmatrix} 1 & 0 & 1 & 0 \\ 0 & 1 & 0 & 1 \\ 0 & 0 & 1 & 0 \\ 0 & 0 & 0 & 1 \end{bmatrix}, \qquad \mathbf{Q} = \begin{bmatrix} 10 & 0 & 0 & 0 \\ 0 & 10 & 0 & 0 \\ 0 & 0 & 1 & 0 \\ 0 & 0 & 0 & 1 \end{bmatrix}.$$

The target randomly changes its direction and is reflected back when it reaches the grid boundary.

A set of $N = 15$ camera sensors monitor the area. The observations are generated using Eq. (2). The observation matrix $\mathbf{H}_i$ and the communication adjacency matrix $\mathbf{A}$ are set as the following

$$\mathbf{H}_i = \begin{bmatrix} 1 & 0 & 0 & 0 \\ 0 & 1 & 0 & 0 \end{bmatrix}, \qquad \mathbf{A} = \begin{bmatrix} 0 & 1 & 0 & 0 & 0 \\ 1 & 0 & 1 & 0 & 0 \\ 0 & 1 & 0 & 1 & 0 \\ 0 & 0 & 1 & 0 & 1 \\ 0 & 0 & 0 & 1 & 0 \end{bmatrix}.$$

If a camera has a measurement, measurement information matrix $\mathbf{B}_i = 0.01\mathbf{I}_2$ is used. Otherwise, $\mathbf{B}_i$ is set to $0\mathbf{I}_2$. The consensus rate parameter ϵ is set to $0.65/\Delta_{max}$ where $\Delta_{max} = 2$.

We conduct the experiment by relaxing the optimality condition where the initial prior states and covariances are different at different nodes. The prior states are initialized by adding Gaussian noise (generated using the corresponding prior covariance matrices) to the initial ground truth states. The initial prior error across different cameras are correlated with correlation coefficient $\rho = 0.5$. The total number of consensus iterations K is varied from 1 to 20 with increments of 1. A total of 15 cameras are used and the camera locations, orientations, network topologies and ground truth tracks are generated randomly.

The results of this experiment are shown in Fig. 3. The simulation results are averaged over 400 independent simulation runs. The mean error (solid lines) and the standard deviation ($\pm 0.2\sigma$ with dotted lines) for different methods are shown using different colors. The results show that ICF achieves near-optimal performance even when the conditions for optimality are not met.

Naivety is a reason for the performance deterioration of the KCF algorithm which is apparent in Fig. 3. As there are 15 cameras and each of these cameras is connected only to two other cameras, there are many naive nodes present in this experiment. The issue of naivety in distributed frameworks was one of the main motivations for the derivation of the GKCF and ICF approach. It can be seen from the results that GKCF and ICF handle the issue of naivety better than the KCF.

ICF also requires low communication bandwidth which is half the required bandwidth of GKCF and comparable with that of KCF. The information sent from each node to a neighbor at each iteration for various methods is shown in Table 1.

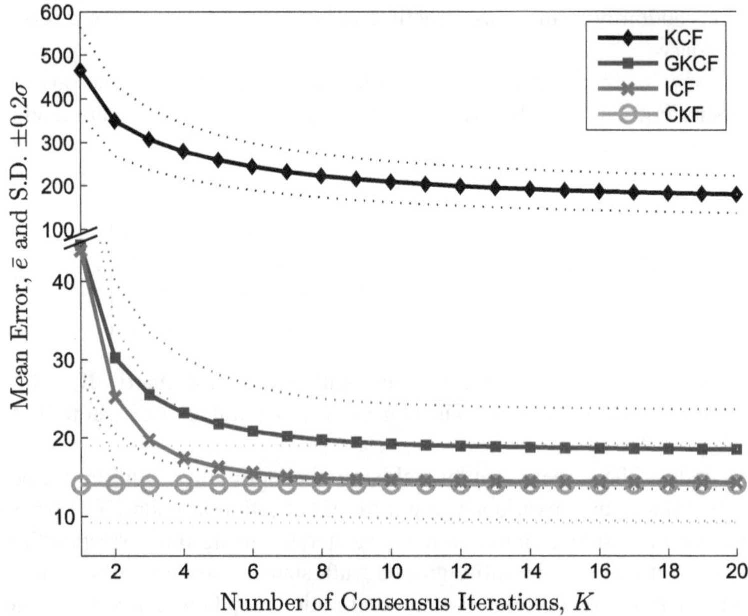

Fig. 3 Performance comparison of different approaches by varying the total number of consensus iterations, K. Each *line* represents the mean error $\bar{e}$ for each method. The *dotted lines* represent the standard deviation ($\pm 0.2\sigma$). The priors at $t = 1$ were set to be different and correlated with $\rho = 0.5$. This figure shows that ICF is robust and achieves near-optimal performance even when the optimality conditions are not met

Table 1 Information sent at each consensus step

Method	Message content
KCF	For 1st consensus step: $\mathbf{u}_i \in \mathcal{R}^p$, $\mathbf{U}_i \in \mathcal{R}^{p \times p}$, $\bar{\mathbf{x}}_i \in \mathcal{R}^p$
	For additional consensus steps: $\bar{\mathbf{x}}_i \in \mathcal{R}^p$
GKCF	$\mathbf{u}_i \in \mathcal{R}^p$, $\mathbf{U}_i \in \mathcal{R}^{p \times p}$, $\bar{\mathbf{x}}_i \in \mathcal{R}^p$, $\mathbf{J}_i \in \mathcal{R}^{p \times p}$
ICF	$\mathbf{v}_i \in \mathcal{R}^p$, $\mathbf{V}_i \in \mathcal{R}^{p \times p}$

3 Collaborative Sensing in Camera Networks

In this section we discuss solution strategies for collaborative sensing in PTZ camera networks [4, 37, 38]. We start by motivating the necessity of a cooperative strategy in an intelligent camera network. The two questions we address are the following—(i) why do we need active cameras (as opposed to having a network of cameras with a fixed set of parameters) and (ii) why does the control strategy need to be cooperative?

The main reason for having a dynamically self-configurable network is that it would be prohibitively expensive to have a static setup that would cater to all possible situations. For example, suppose we need (1) to track one person (possibly non-cooperative) as he walks around an airport terminal, (2) to obtain a high resolution image of him or his specific feature (e.g. face), and (3) to observe other activities going on in the terminal. To achieve this, we will either need to dynamically change the parameters of the cameras where this person is visible or have a setup whereby it would be possible to capture high resolution imagery irrespective of where the person is in the terminal. The second option would be very expensive and a huge waste of resources, both technical and economic. Therefore we need a way to control the cameras based on the sensed data.

The control strategy must necessarily be cooperative because each camera's parameter settings entail certain constraints on other cameras. For example, if a camera zooms in to focus on the face of one particular person, thus narrowing its field of view (FOV), it *risks* losing much of that person and the surroundings; therefore being less useful for target tracking. Another camera can compensate for this by adjusting its parameters. This requires analysis of the video data in a *collaborative network-centric manner*, leading to a cost-effective method to obtain high resolution images for features at dynamically changing locations.

3.1 Overview of Solution Strategy

Each of the cameras in our network has its own embedded target detection module, a distributed tracker that provides an estimate of the state of each target in the scene [3, 48], and finally a distributed camera control mechanism. Given the target detection module, this chapter provides an overview of a distributed solution for the tracking and control problems. Targets will be tracked using measurements from multiple cameras that may not have direct communication between each other. Neighboring cameras will communicate with each other to come to a consensus about the estimated position of each target. Similarly, camera parameters will be assigned based on information exchanged with neighboring cameras within a game-theoretic framework. Our proposed system structure is shown in Fig. 1.

The target detection module takes the image plane measurements and returns the image plane positions associated with specific targets. Through calibration, we know the transformation between image and ground plane coordinates, which can be used for data association. These features along with their measurement error are then passed on to the distributed tracker associated with each target, which then combines the information from neighboring cameras at each imagery time instant to come to a consensus regarding the estimated state and estimated error covariance [3, 48] of each target. This results in each camera having a consensus estimate of the state and error covariance of each target in the region of interest (as explained in Sect. 2).

The camera control module attempts to optimize the scene analysis performance at the next imaging time instant, by selecting the camera parameters expected to result in measurements meeting criteria that are specified by the user, such as minimizing the estimated error covariance of the tracker, maximizing the resolution of the targets, and minimizing the risk of failing to obtain an image of the target. These are represented in a reward versus risk trade-off function. We propose a distributed optimization solution strategy to this problem. Following recent trends in cooperative control and distributed optimization work, this can be represented in a game-theoretic framework.

3.2 Problem Statement and Notation

Consider a region R over which there are $\mathcal{T} = \{T^1, \ldots, T^{N_T}\}$ targets moving independently while being observed by a network of cameras $\mathcal{C} = \{C_1, \ldots, C_{N_C}\}$. The behavior of these camera networks will be determined by a set of criteria specified by the user. Each target T^j, for $j \in \{1, \ldots, N_T\}$, may have different criteria that must be satisfied by the camera network, e.g., we may want to identify T^1 using facial recognition and determine what T^2 is doing using action recognition. In another scenario, we may be interested in observing the scene of a disaster zone to discover and monitor the state of people. Here, we would likely place priority on those who are wounded or in need of aid. This scenario would require targets in R to have varying values of importance and varying requirements over time. The question, is whether we can create a generalized framework for distributed camera networks in which many of the scenarios may be modeled and solved.

In our formulation, the state of the jth target at time step k is defined as $\mathbf{x}^j(k) = (x^j(k), y^j(k), \dot{x}^j(k), \dot{y}^j(k))^T$ where $(x^j(k), y^j(k))$ and $(\dot{x}^j(k), \dot{y}^j(k))$ are the position and velocity of target T^j on the ground plane, respectively. By considering a linear, discrete time, dynamical model

$$\mathbf{x}^j(k+1) = \mathbf{\Phi}^j(k)\mathbf{x}^j(k) + \mathbf{D}^j(k)\gamma^j(k), \tag{39}$$

and nonlinear observation model for each camera C_i,

$$\mathbf{z}_i^j(k) = h_i(\mathbf{x}^j(k)) + \mathbf{v}_i^j(k), \tag{40}$$

the consensus state estimate $\bar{\mathbf{x}}_i^{j-}$ and the error covariance matrices $\mathbf{P}_i^{j-}$ computed as per Sect. 2 are the inputs to the control module.

For camera control, we will design a strategy such that each camera is a rational decision maker, optimizing its own utility function which has been designed so that an increase in the local utility translates to an increase of the global utility function. Let $\mathcal{A}_i$ be the parameter profile that C_i can select from to optimize its own utility function $V_{C_i}(\mathbf{a}_i, \mathbf{a}_{-i})$. Camera $C_i \in \mathcal{C}$ will select its own set of parameters $\mathbf{a}_i \in \mathcal{A}_i$. Our objective is to design these utility functions and appropriate negotiation procedures that lead to mutually agreeable parameter settings of the cameras meeting the global criterion.

The game theoretic interpretation of the distributed optimization in some of the cooperative control work [34, 52] allows for performance analysis in terms of game theoretic results. A well-known concept in game theory is the notion of Nash equilibrium. In the context of our image network problem, it will be defined as a choice of sets of parameters $\mathbf{a}^* = (\mathbf{a}_1^*, \ldots, \mathbf{a}_{N_C}^*)$ such that no sensor could improve its utility further by deviating from $\mathbf{a}^*$. Obviously, $\mathbf{a}^*$ is a function of time since the targets are dynamic and the cameras could also be mobile and capable of panning, tilting and zooming. For our problem, a Nash equilibrium will be reached, at each instant of time, when there is no advantage for a particular camera to choose some other parameter set.

Mathematically, if $\mathbf{a}_{-i}$ denotes the collection of parameter settings for all cameras *except* camera C_i, then $\mathbf{a}^*$ is a *pure Nash equilibrium* if

$$V_{C_i}(\mathbf{a}_i^*, \mathbf{a}_{-i}^*) = \max_{\mathbf{a}_i \in \mathcal{A}_i} V_{C_i}(\mathbf{a}_i, \mathbf{a}_{-i}^*), \forall C_i \in \mathcal{C}. \tag{41}$$

3.3 Choice of Utility Functions and Camera Parameter Assignment

Consider N_C cameras and a time-varying number, $N_T(k)$, of targets with independent, unknown trajectories. Let $\mathbf{p}^j(k)$ denote the position at time k of target T^j for $j = 1, \ldots, N_T$ in the global frame with $\bar{\mathbf{p}}^{j-}(k)$ representing the estimated position and $\mathbf{P}^{j-}$ the error covariance matrix, as obtained from a state estimation algorithm.[1] All cameras have known fixed locations with changeable pan (ρ), tilt (τ), and zoom (ζ) parameters. Let the parameters of camera C_i be denoted by a three vector $\mathbf{a}_i = (\rho_i, \tau_i, \zeta_i)$. Any choice of $\mathbf{a}_i$ yields a field-of-view (FOV) for the resulting image. That image may contain multiple targets and each target may be imaged by multiple cameras. The parameters of all cameras are organized into a vector $\mathbf{a} = [\mathbf{a}_1, \ldots, \mathbf{a}_i, \ldots, \mathbf{a}_{N_c}]$. The vector containing all parameter vectors except those of C_i will be denoted by $\mathbf{a}_{-i}$. We will represent the global imaging utility as $U_I(\mathbf{a} : \bar{\mathbf{p}}^{j-}, \mathbf{P}^{j-})$ and the local imaging utility as $U_{I_i}(\mathbf{a}_i : \mathbf{a}_{-i}, \bar{\mathbf{p}}^{j-}, \mathbf{P}^{j-})$, where $j = 1, \ldots, N_T$. For ease of notation, we will drop a function's dependence on $\bar{\mathbf{p}}^{j-}$ and $\mathbf{P}^{j-}$, unless needed for clarity. Subscripts will be used to represent the physical parameter the utility measures.

3.3.1 Imaging Utility

The global Imaging Utility, represented as $U_I(\cdot)$, should have the following properties.[2]

[1] For this section, we assume that all cameras in the network possess a single consensus state estimate $\bar{\mathbf{x}}^{j-}(k)$ for target T^j with consensus covariance matrix $\mathbf{P}^{j-}$.

[2] Note that the same properties apply to the local imaging utility represented by $U_{I_i}(\cdot)$

- **Enhance image quality**: The utility should increase with image quality. Herein, image quality is defined by two parameters: resolution and aspect angle. Resolution is defined to be the vertical extent of the target on the image, which increases monotonically with focal length and zoom ζ of the imaging camera. Aspect angle

$$\alpha_i^j = \begin{cases} \arccos(\mathbf{o}_{T^j} \cdot \mathbf{o}_{C_i}) & \text{if } \mathbf{o}_{T^j} \cdot \mathbf{o}_{C_i} < 0 \\ 0 & \text{otherwise,} \end{cases} \tag{42}$$

  is defined as the angle between the camera's optical axis and the target's direction of motion and assumes that the target is facing in the direction of its velocity vector.
- **Balanced Risk**: Risk is defined as the probability that the target is outside of the FOV of the cameras that are expected to image it. Risk increases monotonically with focal length, because the FOV decreases as the focal length increases.
- **Continuously differentiable**: In addition to these properties being necessary for the proofs of convergence, they greatly facilitate the numeric optimization process.

Other aspects, like getting images that allow for better activity analysis or object recognition, can also be incorporated into the framework. To understand the issues involved, it is informative to briefly consider the simple case where $N_T = 1$. For this case, if risk was neglected and U_I was defined to be the number of image pixels in the vertical extent of the target, then each camera would maximize its focal length and select its pan and tilt parameters to center on the target. If instead, the utility accounted appropriately for risk, then the set of cameras would adjust their FOV's to cover a sufficiently probable area, with high resolution images being attempted only when sufficient degrees-of-freedom were available.

The risk arises because the imaging utility is dependent on the target position $\mathbf{p}^j(k)$, which is a random variable. The risk is addressed by maximizing the Batesian Imaging Utility

$$V_I(\mathbf{a}) = E_{\mathbf{p}^j}\langle U_I(\mathbf{a} : \bar{\mathbf{p}}^{j-}, \mathbf{P}^{j-})\rangle, \quad j = 1, \ldots, N_T,$$

where the expectation is with respect to each target positions over the FOV of each camera. The probability distribution of each target's location is provided by the target state estimator module.

3.3.2 Tracking Criteria

The purpose of the tracking criteria $U_{T^j}(\mathbf{a} : \bar{\mathbf{p}}^{j-}, \mathbf{P}^{j-}, j = 1, \ldots, N_T)$, where again the dependence on $\bar{\mathbf{p}}^j$ and $\mathbf{P}^{j-}$ will be dropped to simplify notation, is to quantify how well each T^j is expected to be tracked by the camera network for a given parameter vector $\mathbf{a}$. It can be formulated via the Fisher Information matrix to be monotonically increasing with the tracking accuracy of the target. The information matrix is the inverse of posterior error covariance, i.e., $\mathbf{J}^{j+} = \left(\mathbf{P}^{j+}\right)^{-1}$, and is written as

$$(\mathbf{P}^{j+})^{-1} = (\mathbf{P}^{j-})^{-1} + \sum_{l \in (C_i \cup C_i^n)} \mathbf{H}_l^j(\mathbf{a})^\top \mathbf{R}_l^j(\mathbf{a})^{-1} \mathbf{H}_l^j(\mathbf{a}). \tag{43}$$

This matrix provides a measure that represents the accuracy of tracking T^j and can be obtained from the distributed tracker, like the Kalman Consensus Filter [28]. There are multiple ways of defining the tracking utility of T^j, one example being $U_{T^j}(\mathbf{a}) = \min(diag(\mathbf{J}^{j+}))$, as defined in [38]. Modifications to the parameter settings vector $\mathbf{a}$ determines the FOV and measurement function for each camera, see [38], which, in turn, affects the expected tracking accuracy and the Fisher Information. Similar to the Batesian Imaging Utility function, we define a Batesian Tracking Utility function $V_{T^j}(\mathbf{a})$ as the expected value of Tracking Utility $U_{T^j}(\mathbf{a} : \bar{\mathbf{p}}^{j-}, \mathbf{P}^{j-})$ over the FOV of all cameras:

$$V_{T^j}(\mathbf{a}) = E_{\mathbf{p}^j}\left\langle U_{T^j}(\mathbf{a} : \bar{\mathbf{p}}^{j-}, \mathbf{P}^{j-})\right\rangle, **\varphi = \varnothing, \ldots, ,$$

where the expectation is with respect to the target position with each camera's contribution integrated over the FOV of that camera.

3.3.3 Global Utility

The global utility $V_G(\mathbf{a})$ describes the desirability of the settings profile $\mathbf{a}$, given the criteria that must be satisfied by the camera network. Let the target utility associated with target T^j be denoted by $V_j(\mathbf{a})$, and for example, be defined as $V_j(\mathbf{a}) = V_{T^j}(\mathbf{a}) + g(V_{T^j}(\mathbf{a}))V_I(\mathbf{a})$, where $g(V_{T^j}(\mathbf{a}))$ is a function of the tracking utility that returns a non-zero positive value when a user defined tracking criteria is met [38]. Thus, the global utility $V_G(\mathbf{a})$ can then be represented as a function of the importance/priority of each target T^j and its related target utility $V_j(\mathbf{a})$.

$$V_G(\mathbf{a}) = v_1 \cdot V_1(\mathbf{a}) + \cdots + v_{N_T} \cdot V_{N_T}(\mathbf{a}), \tag{44}$$

where v_j denotes the importance of target T^j.

By maximizing the global utility, we are choosing the settings profile that best satisfies the criteria specified for the camera network.

3.3.4 Camera Utility

The global utilities must now be converted into local utilities in order for them to be solved in a distributed fashion. Convergence proofs in game theory [53] require that the local utilities are aligned with the global utility, i.e., a change in the local utility affects the global utility similarly. We achieve this by making the utility of our camera equivalent to its contribution to the global utility, i.e.,

$$V_{C_i}(\mathbf{a}) = V_G(\mathbf{a}) - V_G(\mathbf{a}_{-i}) \tag{45}$$

where the set $\mathbf{a}_{-i}$ is the set of settings profiles excluding the profile for camera C_i. This is known as the Wonderful Life Utility (WLU) [54], and as shown in [53] and applied in [34], leads to a potential game using the global utility as the potential function. This allows us to maximize the global utility through the maximization of the utility of each camera.

3.3.5 Cooperative Game-Theoretic Solution

The dynamic nature of the targets in the region being observed requires that our cameras communicate with each other in order to decide the set of parameters that will result in the optimal global utility. Each camera negotiates with its neighboring cameras to accurately predict the actions of the other cameras before deciding its own action. The overall idea of the proposed negotiation strategy is to use *learning algorithms for multi-player games* [53]. A particularly appealing strategy for this problem is Spatial Adaptive Play (SAP) [55]. This is because it can be implemented with a low computational burden on each camera and leads to an optimal assignment of targets with arbitrarily high probabilities for the local utility described above. Iteration stops if a Nash equilibrium is attained or if the available operation time expires.

3.4 Constrained Optimization Framework

The global utility maximization problem can also be written as a constrained optimization problem where the objective function quantifies the sum of the per camera per target image quality, and the constraint defines the feasible PTZ parameter space that is a lower bound on the information about the estimated position for each target. An example of such a formulation is as follows:

$$\begin{array}{ll} \text{maximize} & V_I(\mathbf{a}) \\ \text{subject to} & \mathbf{V}_{Tr}(\mathbf{a}) \succeq \bar{\mathbf{T}}, \end{array} \tag{46}$$

where $\mathbf{V}_{Tr}(\mathbf{a}) = \left[V_{Tr_1}(\mathbf{a}), \ldots, V_{Tr_{N_T}}(\mathbf{a})\right]^\top$. The constraints can be then augmented into the objective function to form a global *Lagrangian* $L(\boldsymbol{\lambda}, \mathbf{a})$ with

$$L(\boldsymbol{\lambda}, \mathbf{a}) = V_I(\mathbf{a}) + \boldsymbol{\lambda}^\top [\mathbf{V}_{Tr}(\mathbf{a}) - \bar{\mathbf{T}}], \tag{47}$$

where $L : (\boldsymbol{\lambda}, \mathbf{a}) \mapsto \Re$, and $\boldsymbol{\lambda} = [\lambda^1, \ldots, \lambda^{N_T}]$ is the Lagrange multiplier vector, where each element λ^j is the Lagrange multiplier for the inequality constraint associated with the jth target. Thus, to find the optimal primal-dual pair of

Fig. 4 Effect of distributed PTZ camera control. *Top row* shows images from four (non-optimized) cameras covering the entire area (*last column*) yielding low resolution target images, making analysis difficult. *Bottom row* shows images from optimized PTZ settings yielding high-resolution for two targets, while still keeping the other targets in view. The results are from a prototype system developed based on our preliminary work and which is running in the UCR camera network facility

solutions $(\mathbf{a}^*, \boldsymbol{\lambda}^*)$ through a central controller, the global unconstrained problem given by the Lagrangian in Eq. (47) would have to be solved.

3.5 Experimental Results

To evaluate the proposed camera control algorithm, we show results on a real-life camera network by considering an example performance criteria for scene analysis, i.e., Area Coverage. The sensor network consists of 4 PTZ IP cameras with a resolution of 320×240 pixels and 12x optical zoom. In this experiment for each camera, the pan and tilt parameters are roughly separated by 15° with 2 settings for zoom. We use this setting to show the performance of our approach with an increasing number of cameras removed from the area coverage optimization. Figure 4 shows the advantage of controlling the cameras to acquire high resolution images of the targets (bottom row) as opposed to the case where the details on the targets are not viewable when the entire area is covered by a pre-defined camera arrangement (top row).

References

1. Olfati-Saber, R.: Kalman-consensus filter Optimality, stability, and performance. In: IEEE Conference on Decision and Control (2009)
2. Kamal, A.T., Ding, C., Song, B., Farrell, J.A., Roy-Chowdhury, A.K.: A generalized Kalman consensus filter for wide-area video networks. In: IEEE Conference on Decision and Control (2011)
3. Kamal, A.T., Farrell, J.A., Roy-Chowdhury, A.K.: Information weighted consensus. In: IEEE Conference on Decision and Control (2012)

4. Ding, C., Song, B., Morye, A.A., Farrell, J.A., Roy-Chowdhury, A.K.: Collaborative sensing in a distributed PTZ camera network. IEEE Trans. Image Process **21**(7), 3282–3295 (2012)
5. Davis, J.W.: Camera control and geo-registration for video. In: Bhanu, B., Ravishankar, C., Roy-Chowdhury, A.K., Aghajan, H., Terzopoulos, D. (eds.) Distributed Video Sensor Networks. Springer, Berlin (2011)
6. Markis, D., Ellis, T., Black, J.: Bridging the gap between cameras. In: IEEE Conference on Computer Vision and Pattern Recognition (2004)
7. Tieu, K., Dalley, G., Grimson, W.E.L.: Inference of non-overlapping camera network topology by measuring statistical dependence. In: IEEE International Conference on Computer Vision (2005)
8. Song, B., Roy-Chowdhury, A.K.: Stochastic adaptive tracking in a camera network. In: IEEE International Conference on Computer Vision (2007)
9. Rahimi, A., Darrell, T.: Simultaneous calibration and tracking with a network of non-overlapping sensors. In: IEEE Conference on Computer Vision and Pattern Recognition (2004)
10. Khan, S., Javed, O., Rasheed, Z., Shah, M.: Human tracking in multiple cameras. In: IEEE International Conference on Computer Vision, vol. 1, pp. 331–336 (2001)
11. Alahi, A., Marimon, D., Bierlaire, M., Kunt, M.: A master-slave approach for object detection and matching with fixed and mobile cameras. In: International Conference on Image Processing (2008)
12. Ermis, E., Saligrama, V., Jodoin, P., Konrad, J.: Abnormal behavior detection and behavior matching for networked cameras. In: IEEE/ACM International Conference on Distributed Smart Cameras (2008)
13. Javed, O., Khan, S., Rasheed, Z., Shah, M.: Camera handoff: tracking in multiple uncalibrated stationary cameras. In: IEEE Workshop on Human Motion, Dec. 2000
14. Li, Y., Bhanu, B.: Utility based dynamic camera assignment and hand-off in a video network. In: IEEE/ACM International Conference on Distributed Smart Cameras (2008)
15. Stancil, B., Zhang, C., Chen, T.: Active multicamera networks: from rendering to surveillance. IEEE J. Sel. Top. Signal Process. (Special Issue Distrib. Process. Vis. Netw.) **2**, 597–605 (2008)
16. Zhao, J., Cheung, S.C., Nguyen, T.: Optimal camera network configurations for visual tagging. IEEE J. Sel. Top. Signal Process. **2**(4), 464–479 (2008)
17. Erdem, U.M., Sclaroff, S.: Automated camera layout to satisfy task-specific and floor plan-specific coverage requirements. Comput. Vis. Image Underst. **103**(3), 156–169 (2006)
18. Du, W., Piater, J.: Multi-camera people tracking by collaborative particle filters and principal axis-based integration. In: Asian Conference on Computer Vision (2007)
19. Khan, S., Shah, M.: A multiview approach to tracking people in crowded scenes using a planar homography constraint. In: European Conference on Computer Vision (2006)
20. Devarajan, D., Cheng, Z., Radke, R.J.: Calibrating distributed camera networks. Proc. IEEE (Special Issue Distrib. Smart Cameras) **96**(10), 1625–1639 (2008)
21. Sankaranarayanan, A.C., Veeraraghavan, A., Chellappa, R.: Object detection tracking and recognition for multiple smart cameras. Proc. IEEE (Special Issue Distrib. Smart Cameras) **96**(10), 1606–1624 (2008)
22. Wu, C., Aghajan, H.: Real-time human pose estimation: a case study in algorithm design for smart camera networks. Proc. IEEE (Special Issue Distrib. Smart Cameras) **96**(10), 1715–1732 (2008)
23. Olfati-Saber, R., Fax, J.A., Murray, R.M.: Consensus and cooperation in networked multi-agent systems. Proc. IEEE **95**(1), 215–233 (2007)
24. Tron, R., Vidal, R.: Distributed algorithms for camera sensor networks. IEEE Signal Process. Mag. **3**, 32–45 (2011)
25. Tron, R., Vidal, R.: Distributed computer vision algorithms through distributed averaging. In: IEEE Conference on Computer Vision and Pattern Recognition (2011)
26. Kamal, A.T., Farrell, J.A., Roy-Chowdhury, A.: Optimal distributed maximum a posteriori estimation. In: International Conference on Image Processing (2012)

27. Taj, M., Cavallaro, A.: Distributed and decentralized multicamera tracking. IEEE Signal Process. Mag. **3**, 46–58 (2011)
28. Song, B., Kamal, A.T., Soto, C., Ding, C., Farrell, J.A., Roy-Chowdhury, A.K.: Tracking and activity recognition through consensus in distribution camera network. IEEE Trans. Image Process. **19**(10), 2564–2579 (2010)
29. Kamal, A.T., Song, B., Roy-Chowdhury, A.K.: Belief consensus for distributed action recognition. In: International Conference on Image Processing, pp. 141–144, Sept. 2011
30. Song, B., Ding, C., Kamal, A.T., Farrell, J.A., Roy-Chowdhury, A.K.: Distributed camera networks: integrated sensing and analysis for wide area scene understanding. IEEE Signal Process. Mag. **3**, 20–31 (2011)
31. Micheloni, C., Rinner, B., Foresti, G.L.: Video analysis in Pan-Tilt-Zoom camera networks. IEEE Signal Process. Mag. **27**(5), 78–90 (2010)
32. Micheloni, C., Foresti, G., Snidaro, L.: A network of cooperative cameras for visual-surveillance. IEE Proc. Visual Image Signal Process. **152**(2), 205–212 (2005)
33. Qureshi, F.Z., Terzopoulos, D.: Surveillance in virtual reality: system design and multi-camera control. In: IEEE Conference on Computer Vision and Pattern Recognition (2007)
34. Arslan, G., Marden, J., Shamma, J.: Autonomous vehicle-target assignment: a game-theoretical formulation. ASME J. Dyn. Syst. Meas. Control **129**(5), 584–596 (2007)
35. Song, B., Soto, C., Roy-Chowdhury, A.K., Farrell, J.A.: Decentralized camera network control using game theory. In: Workshop on Smart Camera and Visual Sensor Networks at ICDSC (2008)
36. Soto, C., Song, B., Roy-Chowdhury, A.K.: Distributed multi-target tracking in a self-configuring camera network. In: IEEE Conference on Computer Vision and Pattern Recognition (2009)
37. Ding, C., Morye, A.A., Farrell, J.A., Roy-Chowdhury, A.K.: Coordinated sensing and tracking for mobile camera platforms. In: American Control Conference (2012)
38. Morye, A.A., Ding, C., Song, B., Roy-Chowdhury, A.K., Farrell, J.A.: Optimized imaging and target tracking within a distributed camera network. In: American Control Conference (2011)
39. Blake, A., Yuille, A. (eds): Active Vision. MIT Press, Cambridge (1992)
40. Gu, G., Chandler, R., Schumacher, C., Sparks, A., Pachter, M.: Optimum cooperative UAV sensing based on Cramer-Rao bound. In: American Control Conference, June 2005)
41. Frew, E.W.: Sensitivity of cooperative geolocalization to orbit coordination. In: AIAA Guidance, Navigation, and Control Conference, Hilton Head, SC, Aug. 2007
42. Kingston, D.B.: Decentralized control of multiple UAVs for perimeter and target surveillance. Ph.D. Dissertation, Brigham Young University, Provo, Utah, Dec. 2007
43. Quintero, S., Papi, F., Klein, D., Chisci, L., Hespanha, J.: Optimal UAV coordination for target tracking using dynamic programming. In: IEEE Conference on Decision and Control, Dec. 2010
44. Stachura, M., Carfang, A., Frew, E.W.: Cooperative target tracking with a communication limited active sensor network. In: International Workshop on Robotic Wireless Sensor Networks (2009)
45. Olfati-Saber, R., Murray, R.M.: Consensus problems in networks of agents with switching topology and time-delays. IEEE Trans. Autom. Control **49**(9), 1520–1533 (2004)
46. Moreau, L.: Stability of multiagent systems with time-dependent communication links. IEEE Trans. Autom. Control **50**(2), 169–182 (2005)
47. Chen, Y., Tron, R., Terzis, A., Vidal, R.: Corrective consensus: converging to the exact average. In: IEEE Conference on Decision and Control, pp. 1221–1228, Dec. 2010
48. Olfati-Saber, R., Sandell, N.F.: Distributed tracking in sensor networks with limited sensing range. In: Proceedings of the American Control Conference, Seattle, WA, USA, pp. 3157–3162, June 2008
49. Ren, W., Beard, A.W., Kingston, D.B.: Multi-agent Kalman consensus with relative uncertainty. In: American Control Conference (2005)

50. Alighanbari, M., How, J.P.: An unbiased Kalman consensus algorithm. In: American Control Conference (2006)
51. Kalman, R.: A new approach to linear filtering and prediction problems. Trans. ASME (J. Basic Eng.) **82**(Series D), 35–45 (1960)
52. Marden, J., Arslan, G., Shamma, J.: Cooperative control and potential games. IEEE Trans. Syst. Man Cybern. Part B: Cybernet. **39**(6), 1393–1407 (2009)
53. Monderer, D., Shapley, L.S.: Potential games. Games Econ. Behav. **14**(1), 124–143 (1996)
54. Wolpert, D., Tumor, K.: A survey of collectives, In: Tumer, K., Wolpert, D. (eds.) Collectives and the Design of Complex Systems. Springer, New York (2004)
55. Young, H.P.: Individual Strategy and Social Structure: An Evolutionary Theory of Institutions. Princeton University Press, Princeton (1998)

Index

Augment Vis Real (2014) 6: 235–236
DOI: 10.1007/978-3-642-37841-6